The Blue Book

Declassified
Unidentified Flying Object (UFO)
Reports and Documents

Concise Reference
Volume 1

The Blue Book

Declassified
Unidentified Flying Object (UFO)
Reports and Documents

Concise Reference
Volume 1

The Blue Book: Declassified Unidentified Flying Object (UFO) Reports and Documents Concise Reference Volume 1

Oak Hill Books
PO Box 461
Willow Springs, NC 27592

Author / Editor: Mindy J. Allport-Settle

Published by Oak Hill Books

Printed in the United States of America.

ISBN 1-9372580-9-2
ISBN-13 978-1-937258-09-2

Contents

National Security Agency FOIA Civil Action Documents

Overview

About this Book

Declassification of UFO documents began at the country level with the United Kingdom in 2008. Russia, Ecuador, the United States, France, Denmark, Brazil, Sweden, Canada and many other countries followed that decision during 2009 making public all of the material they had in their possession.[1]

Volume 1 of *the Blue Book* includes all of the documents made available to the public from the United States National Security Agency. This book includes all of the complete documents exactly as they were released by the government with no omissions or additions.

Included Documents and Features

National Security Agency FOIA Civil Action Documents

1. * In Camera Affidavit of Eugene F. Yeates - Citizens Against Unidentified Flying Objects Secrecy v. NSA (Civil Action No. 80-1562)
2. * Communications Intelligence (COMINT) reports
3. * Affidavit of Eugene F. Yeates - Citizens Against Unidentified Flying Objects Secrecy v. NSA (Civil Action No. 80-1562)
4. "U.F.O. Files: The Untold Story" by Patrick Huyghe
5. "Govt.'s Super-Secret Security Agency Warns: Take UFOs Seriously or Be Prepared for Sneak Invasion by Space Aliens" by Thomas L. Muldoon, National Enquirer
6. "The U.S. Government and the Iran Case"- IUR Report
7. "Report of U.F.O. Crash in '47 Called False by Science Panel"- The New York Times
8. "What the U.S. Government Knows About UnidentifiedFlying Objects" by Peter Gersten, Frontiers of Science
9. "Is the CIA Stonewalling?" by Richard Hall

[1] http://www.nsa.gov/public_info/declass/ufo/

10. "The Government and UFOs"
11. "UFO As Advanced Technology"
12. "French Government UFO Study"
13. Communications Intelligence (COMINT) report - "XXXXX Unidentified Flying Objects"
14. COMNAVSECGRU visual sighting report
15. "Report Bibliography" issued by the Defense Documentation Center
16. "Now You See It, Now You Don't" by Captain Henry S. Shields, HQ USAFE/INOMP
17. Joint Chiefs of Staff report concerning the sighting of a UFO in Iran on 19 September 1976
18. Exploitation Report - A Fragment, Metal Recovered in the Republic of the Congo
19. Navy report - Country: Cuba, Subject: UFO
20. Routing and Transmittal Slip from Louis E. Foster, DIA with a DIA summary sheet and US Defense Attache report on the sighting of a UFO in Iran on 19 September 1976
21. Department of State AIRGRAMs and Reports:
    - Subject: Flying Saucers Are a Myth
    - Subject: Tunisian Firefall
    - Subject: Celestial Body Observed over Antarctica
    - Subject: Political Report for August 1965
    - Subject: Unidentified Flying Objects
    - Subject: Sighting of Object Possibly Originating from Space Vehicle
    - Subject: Unidentified Flying Objects
    - Subject: Unidentified Flying Objects Reported Over Angola
    - Subject: Report of Unidentified Object Which Fell Near Municipio of General Teran, N.L.
30. American Embassy Tunis report - Subject: Unidentified Flying Objects
31. Report from Air Force Special Security Office - Subject: (Unclassified) UFOB NEAC Area

32. General Accounting Office Report to the Honorable Steven H. Schiff, House of Representatives - Government Records - Results of a Search for Records Concerning the 1947 Crash Near Roswell, New Mexico
33. * Memorandum and Order - Citizens Against UFO Secrecy v. National Security Agency (Civil Action No. 80-1562)
34. "Communication With Extraterrestrial Intelligence" by Lambros D. Callimahos
35. * "UFO Hypothesis and Survival Questions"
36. * "UFO's and the Intelligence Community Blind Spot to Surprise or Deceptive Data"
37. Memorandum for the Record - Subject: Information Request Solicitation
38. United States Air Force Fact Sheet 95-03 - "Unidentified Flying Objects and Air Force Project Blue Book"
39. Report of Air Force Research Regarding the "Roswell Incident"
40. JANAP 146
41. Key to the Extraterrestrial Messages by H. Campaigne
42. Extraterrestrial Intelligence by Howard H. Campaigne

Introduction

UFO Documents Index

The documents listed on this page[2] were located in response to the numerous requests received by NSA on the subject of Unidentified Flying Objects (UFO). In 1980, NSA was involved in Civil Action No. 80-1562, "Citizens Against Unidentified Flying Objects Secrecy v. National Security Agency." Documents related to that ligitation are marked with "*". "XXXXX" has been inserted in a title if a portion of the title has been deleted prior to release. To select a document click on the document title, and wait for the PDF version to be downloaded to your local viewer. Approximate file sizes are given after each selection for user convenience.

1. * In Camera Affidavit of Eugene F. Yeates - Citizens Against Unidentified Flying Objects Secrecy v. NSA (Civil Action No. 80-1562)
2. * Communications Intelligence (COMINT) reports
3. * Affidavit of Eugene F. Yeates - Citizens Against Unidentified Flying Objects Secrecy v. NSA (Civil Action No. 80-1562)
4. "U.F.O. Files: The Untold Story" by Patrick Huyghe
5. "Govt.'s Super-Secret Security Agency Warns: Take UFOs Seriously or Be Prepared for Sneak Invasion by Space Aliens" by Thomas L. Muldoon, National Enquirer
6. "The U.S. Government and the Iran Case"- IUR Report
7. "Report of U.F.O. Crash in '47 Called False by Science Panel"- The New York Times
8. "What the U.S. Government Knows About UnidentifiedFlying Objects" by Peter Gersten, Frontiers of Science
9. "Is the CIA Stonewalling?" by Richard Hall
10. "The Government and UFOs"

---

[2] Available from the National Security Agency website at: http://www.nsa.gov/public_info/declass/ufo/

11. "UFO As Advanced Technology"
12. "French Government UFO Study"
13. Communications Intelligence (COMINT) report - "XXXXX Unidentified Flying Objects"
14. COMNAVSECGRU visual sighting report
15. "Report Bibliography" issued by the Defense Documentation Center
16. "Now You See It, Now You Don't" by Captain Henry S. Shields, HQ USAFE/INOMP
17. Joint Chiefs of Staff report concerning the sighting of a UFO in Iran on 19 September 1976
18. Exploitation Report - A Fragment, Metal Recovered in the Republic of the Congo
19. Navy report - Country: Cuba, Subject: UFO
20. Routing and Transmittal Slip from Louis E. Foster, DIA with a DIA summary sheet and US Defense Attache report on the sighting of a UFO in Iran on 19 September 1976
21. Department of State AIRGRAM - Subject: Flying Saucers Are a Myth
22. Department of State report - Subject: Tunisian Firefall
23. Department of State AIRGRAM - Subject: Celestial Body Observed over Antarctica
24. Department of State AIRGRAM - Subject: Political Report for August 1965
25. Department of State AIRGRAM - Subject: Unidentified Flying Objects
26. Department of State AIRGRAM - Subject: Sighting of Object Possibly Originating from Space Vehicle
27. Department of State AIRGRAM - Subject: Unidentified Flying Objects
28. Department of State AIRGRAM - Subject: Unidentified Flying Objects Reported Over Angola
29. Department of State AIRGRAM - Subject: Report of Unidentified Object Which Fell Near Municipio of General Teran, N.L.
30. American Embassy Tunis report - Subject: Unidentified Flying Objects

31. Report from Air Force Special Security Office - Subject: (Unclassified) UFOB NEAC Area
32. General Accounting Office Report to the Honorable Steven H. Schiff, House of Representatives - Government Records - Results of a Search for Records Concerning the 1947 Crash Near Roswell, New Mexico
33. * Memorandum and Order - Citizens Against UFO Secrecy v. National Security Agency (Civil Action No. 80-1562)
34. "Communication With Extraterrestrial Intelligence" by Lambros D. Callimahos
35. * "UFO Hypothesis and Survival Questions"
36. * "UFO's and the Intelligence Community Blind Spot to Surprise or Deceptive Data"
37. Memorandum for the Record - Subject: Information Request Solicitation
38. United States Air Force Fact Sheet 95-03 - "Unidentified Flying Objects and Air Force Project Blue Book"
39. Report of Air Force Research Regarding the "Roswell Incident"
40. JANAP 146
41. Key to the Extraterrestrial Messages by H. Campaigne
42. Extraterrestrial Intelligence by Howard H. Campaigne

Please Note: These historical documents are images of formerly classified carbon paper and reports that have been declassified. Due to the age and poor quality of some of the images, some documents may be difficult to read. In accordance with Section 504 of the Rehabilitation Act of 1973, as amended, individuals may request that the government provide auxiliary aids or services to ensure effective communication of the substance of the documents. For such requests, please contact the Public Affairs Office at 301-688-6524.

UFO Related Information (No Records Exist)

The following terms have been searched in response to requests for information on Unidentified Flying Objects (UFOs) and Paranormal Events, but no responsive material has been located.

---A---

Aegis

AEO (aka Anomalous Aerial Objects)

AF-19741

AF44-29125

AGOR-8

Airgram A-894

ALF

Alien Abduction(s)

Alien Autopsy

Alien Life Form

Animal Mutilations

Anomalous Aerial Objects (aka AEO)

Apollo 8, 11, 12, 14, 15, 16

Aquarius

Area 51 (aka Groom Dry Lake)

Artic(h)oke

Aurora

---B---

Bando

Bermuda Triangle

Big Bird

Black Horse

Bluebird

Bluebolt

Blue Fly

Blue Gemini

Blue Light

Blue Paper

Brazel Ranch

Brookfield Township

---C---

Cattle Mutilations

CETI

Cetus

Corona

Crop Circles

---D---

Dark X

Declination

Deliverance

Delphi

Dreamland

---E---

Eldridge

Electronic Voice Phenomenon

Element 118

Eltanin

ET

ETI

Excalibur

Extraterrestrial, Extraterrestrial Being, ETB

Extraterrestrial Biological Entity, EBE

Extraterrestrial Craft, ETC

Extraterrestrial Research Agency, Exterra

Extraterrestrial Signals

---F---

F-89c

Fastwalkers

Faded Giant

Fire Wheel

Flying Disc, Flying Disk

Flying Saucer

Foo Fighters

Freedom Ridge

---G---

Gabriel

Gemini 4,7, 8, 11, 12

Gleem

Groom Dry Lake (aka Area 51)

Group of 12

Group 13

Grudge 13

---H---

Hacker

Hanger 18

Humanoid Piloted Alien Craft (HPAC)

---I---

Identified Alien Craft (IAC)

Interdimensional Craft

IPU Interplanetary Phenomenon Unit

---J---

Jason Society

JM-8

Joshua

---K---

Kecksburg

Kinross

---L---

Landbroker

La Paz

Liberty Township

Livestock Mutilations

Looking Glass

---M---

MAJCOM

Martin Mariner

Maybell

Men in Black

Menzel

Midnight

Missing Time

MJ12

MK Delta

MK Search

MK Ultra

Montauk

Moondust

MUROC

---N---

Nellis Air Force Base

NEPA

Nighthawk

NORAD Unknown Track Reports

Non-extraterrestrial

---O---

Oasis

Operation Majority

OPNAC Team

Orion

Overview

---P---

Pando

Papoose Lake (aka S4)

Philadelphia Experiment/Project

PI-40

Plato

Pluto, Project Pluto

Pounce

PPS-100

Preserve Destiny

PROJECT BLUETEAM

PROJECT DELTA

PROJECT GARNET

PROJECT NRO

PROJECT URBANE 25

---R---

Rainbow

Ranchhand

Rand

Red Light

Red Team

Rendlesham Forest, England

Right Ascension

Risk

Rockefeller Documents

---S---

S-4

Saint

San Augustin

Sanctum

Saucer

Search Plane Lost While Looking for Navy Flight 19

Semjase

Senior Citizen

Serial No. 51-5853A

SETI

Sidekick

Sigma

Silver Bug

SOM1-01 MJ12

Sorocco

Snow Bird

Star System Zeta Reticuli

Stigmata

Stork

---T---

TAV, TransAtmospheric Vehicle

Traux Field, Wisconsin

Trumbull City

Twinkle

---U---

UAO, Unidentified Aerial Object

UFA, Unidentified Flying Aerodyne

UFO Global Grid System

UFO Recoveries

Ultra Terrestrials

Umbrella

Unconventional Aircraft

Uncorrelated Targets

Unidentified Aerial Object

Unidentified Flying Aerodyne

Unified Field, Unifield

Unit 731

URBANE 25

---W---

Woodbridge, England

Whirlwind

White Sands

White Shot

---Y---

Youngstown Air Force Base

---Z---

Zeta Reticuli

---4---

433rd Fighter Interceptor Squadron

---5---

5412 Group

This page intentionally left blank.

Citizens Against Unidentified Flying Objects Secrecy v. NSA

(Civil Action No. 80-1562)

*** Affidavit of Eugene F. Yeates**

This page intentionally left blank.

c. 10 Oct 80.

UNITED STATES DISTRICT COURT
FOR THE DISTRICT OF COLUMBIA

CITIZENS AGAINST UNIDENTIFIED FLYING OBJECTS SECRECY Plaintiff v. NATIONAL SECURITY AGENCY Defendant	CIVIL ACTION NO. 80-1562

AFFIDAVIT OF EUGENE F. YEATES

State Of Maryland)
) ss.
County Of Anne Arundel)

Eugene F. Yeates, being duly sworn, deposes and says:

1. I am the Chief, Office of Policy, of the National Security Agency (NSA). As Chief, Office of Policy, I am responsible for processing all initial requests made pursuant to the Freedom of Information Act (FOIA) for NSA records. I have read and am familiar with the allegations contained in the Complaint in this case. The statements herein are based upon my knowledge, upon my personal review of information available to me in my official capacity, and upon conclusions reached in accordance therewith.

2. I submit this affidavit regarding all records that have been located by the NSA pursuant to the plaintiff's FOIA requests, in support of the Agency's Motion for Summary Judgment. To supplement this testimony, I will execute an additional affidavit, classified TOP SECRET, to provide highly classified information which is not available in the public realm and which specifically identifies the NSA records being withheld from the plaintiff for the Court's <u>in camera</u> consideration.

THE RELEVANT DOCUMENTS

3. This Agency received four separate referrals (described below) from the Central Intelligence Agency (CIA) concerning NSA documents located by that Agency when it was engaged in FOIA litigation concerning information pertaining to UFOs. (Ground Saucer Watch v. CIA, D.D.C. #78-859.) The first was on November 9, 1978 when this Agency received copies of fifteen Agency documents located by CIA. The NSA also received two other referrals from CIA on December 4, 1978 which provided three additional NSA documents. Subsequently by letter dated December 14, 1978, attorney for the plaintiff, Peter A. Gersten (who was the attorney of record in the CIA litigation) was informed by CIA of the referral of the documents to this Agency. Mr. Gersten then filed a request with NSA for the eighteen documents in his own name on December 22, 1978. In a letter dated January 9, 1979, Roy R. Banner, Chief, Policy Staff, advised Mr. Gersten that the NSA records that had been referred to this Agency by CIA were exempt from release under 5 U.S.C.§ 552(b)(1) because they are classified in their entirety, and exempt from release under 5 U.S.C. § 552(b)(3) because they are exempt from release by 18 U.S.C. § 798, 50 U.S.C. § 403(d)(3) and Section 6 of Public Law 86-36 (50 U.S.C. 402 (note)). (A copy of this letter is attached to the Complaint as Exhibit C. A true and correct copy is also attached to this affidavit as Exhibit 1.)

4. By letter dated January 29, 1979 Mr. Gersten appealed the denial of the documents referred by the CIA. (A copy of this letter is attached to the Complaint as Exhibit D. A true and correct copy of this letter (less enclosures) is also attached to this affidavit as Exhibit 2.) On March 2, 1979 the NSA Freedom of Information Act/ Privacy Act Appeals Authority affirmed the denial of the request for release of the information

on the same grounds, i.e., that the information is classified in its entirety and therefore exempt from release under 5 U.S.C. § 552(b)(1) and that the information is exempt from release under 5 U.S.C. § 552(b)(3) for the reason that other statutes prevent its disclosure, to wit: 18 U.S.C. § 798, 50 U.S.C. § 403(d)(3) and Section 6 of Public Law 86-36. (A copy of this letter is attached to the Complaint as Exhibit E. A true and correct copy is also attached to this affidavit as Exhibit 3.)

5. On February 16, 1979 plaintiff in the instant case, by counsel, filed an FOIA request with NSA for all documents in possession or under the control of NSA relating to UFOs and the UFO phenomena.(A copy of this letter is attached to the Complaint as Exhibit F. A copy is also attached hereto and marked Exhibit 4.) By letter dated January 10, 1980, Mr. Banner released two documents within the scope of the request; deletions of information exempt under the provisions of 5 U.S.C. § 552(b)(1) and (b)(3) were taken prior to that release. Plaintiff was also advised in this letter that other NSA documents existed but were exempt from release under 5 U.S.C. § 552(b)(1) because they are classified in their entirety; exempt from release under 5 U.S.C. § 552(b)(3) because disclosure of them was prohibited by 18 U.S.C. §798, 50 U.S.C. § 403(d)(3) and Public Law 86-36; exempt from disclosure under 5 U.S.C. § 552(b)(5) because they are intra agency memoranda, or letters which would not be available by law to a party other than an agency in litigation with this Agency; or exempt under 5 U.S.C. § 552(b)(6) because they are to personnel or similar files the disclosure of which would constitute a clearly unwarranted invasion of personal privacy. (A copy of this letter is attached to the Complaint as Exhibit G. A true and correct copy is also attached to this affidavit as Exhibit 5.) The January 10 letter also advised the plaintiff that certain information,

originating with other Federal agencies or components would be referred to the originating agency. A total of seventy-nine documents were referred to various agencies for their direct response to plaintiff.

6. An additional NSA document was referred to the NSA by the CIA on July 27, 1979. (The fourth and final CIA referral.) In a letter dated December 19, 1979, Mr. Banner responded to the referral by again advising Mr. Gersten that this record is exempt from release under 5 U.S.C. § 552(b)(1) because it was classified in its entirety, and exempt from release under 5 U.S.C. § 552(b)(3) because disclosure of it was prohibited by 18 U.S.C. § 798, 50 U.S.C. § 403(d)(3) and Public Law 86-36. (A true and correct copy of this letter is attached as Exhibit 6.)

7. By letter dated January 23, 1980, plaintiff appealed the denial of the information referred by CIA and the subsequent denial of his general FOIA request of February 16, 1979. By reply letter dated March 24, 1980, the NSA Freedom of Information Act/Privacy Act Appeals Authority affirmed the denial of plaintiff's request for release of information on the same grounds as cited above. (A copy of this letter is attached to the Complaint as Exhibit K. A true and correct copy is also attached to this affidavit as Exhibit 7.)

8. Finally, a portion of a document which is responsive to the plaintiff's request was recently declassified and released to counsel for the plaintiff by letter dated September 15, 1980. (A true and correct copy of this letter less the enclosure is attached as Exhibit 8.)

THE STATUS AND MISSION OF NSA

9. NSA was established by Presidential Directive in October 1952 as a separately organized Agency within the Department of Defense under the direction, authority, and control of the Secretary of Defense, who was designated by the President as Executive Agent of the Government for conducting

the communications security activities and signals intelligence activities of the United States. A primary signals intelligence (SIGINT) mission of NSA is to intercept radio communications sent to or from foreign governments in order to obtain foreign intelligence information necessary to the national defense, national security, or the conduct of foreign affairs. The intelligence information derived from the intercept of foreign communications is called communications intelligence (COMINT). Of the NSA records at issue in this case, only four are not COMINT reports and will be addressed separately. All of the remaining records sought in this civil action are the product of signals intelligence activities.

10. These COMINT activities include the targeting of foreign governments' communications both within their borders and to and from communicants abroad involving the use of their own radio transmitter and receiver facilities not available for public use. Such communications links are known as "government net" communications. A foreign government may use other means for sending and receiving international radio communications in addition to or instead of its own government facilities. (International radio communications as used here includes communications passed at least in part by wire.) This involves the foreign government's use of the facilities of an international communications common-carrier which is also available for use by the public. Such common-access carriers supply the means by which more than half the encrypted and plain text radio communications of foreign governments, foreign organizations, and their representatives are carried.

11. It is common knowledge that the total volume of radio signals transmitted on a given day is vast. It is also generally known that radio transmissions can be received by anyone

5

X

operating the proper receiving equipment in the right place at the right time. Thus, the fact that NSA can intercept radio communications is generally known. So, too, is it known to foreign officials that such interception of radio communications is a primary mission of the NSA. Foreign officials may be expected to know, also, that NSA cannot possibly intercept even a significant percentage of all such communications, especially taking account of the fact that NSA's activities involve worldwide communications, not solely those having a United States terminal. The number is simply too vast to be handled with any reasonable amount of personnel and equipment. Moreover, the cost and effort of such interception would be disproportionate to the intelligence value of the results.

12. Instead -- as NSA's foreign intelligence targets presumably know well -- NSA must focus its interception activities on those particular communications lines, channels, links or systems which yield the highest proportion of useful foreign intelligence information. What foreign government officials do not know, however, is which of the vast number of radio communications NSA attempts to intercept, which are intercepted, and, of those that are intercepted, which yield to NSA processing methods and techniques. It is the protection of this critical information that is at the heart of the instant case.

13. The continued efficacy of NSA's vital intelligence activities requires that the lines, channels, links and systems actually monitored remain unidentified. If a foreign government obtains sufficient reason to suspect that NSA is able to intercept and process that government's radio communications, that government would be expected to take immediate steps to defeat that capability. This can be accomplished in a number of ways. A foreign government might shift to communications

6

links the U.S. cannot intercept. It may also choose to use alternate methods of communications. The foreign governments may possess the technical capability to either upgrade or initiate cryptography to secure its communications. Finally, a communication channel believed to be targeted by NSA can be used by a foreign government to pass misleading information.

14. If a foreign power is successful in defeating an interception operation, all of the intelligence from that source is lost unless and until NSA can establish a new and equivalent intercept. The risk involved is great. The information produced by NSA includes political, economic, scientific and military data which is of immeasurable value to the President, the Secretary of Defense, the Secretary of State and other policymakers. Obviously, if a COMINT source used by the Agency becomes unavailable, policymakers must operate without the information that source produced. Sometimes it is impossible to establish a new and equivalent intercept and the source is lost permanently. Those losses are not only extremely harmful to the national security but also impose a heavy burden on the limited resources of NSA which must attempt to recover the old source or establish an equivalent source of information.

15. Even after targeting only a small proportion of all available electromagnetic communications for interception, the number of messages intercepted is extremely large. NSA thus is faced with a considerable task in selecting out those messages that will be reviewed for possible intelligence interest. The manner in which NSA does this selection and the degree of reliability and success its methods enjoy are subjects about which virtually no authoritative information has ever been released to the public. Information about these subjects would enable foreign observers to further assess, and thus take

steps to defeat, the capabilities of NSA's intelligence gathering techniques.

RELEASE OF COMINT REPORTS

16. The COMINT reports being withheld from the plaintiff are all based on intercepted foreign communications. The disclosure of these records would identify the communications that had been successfully intercepted and processed for intelligence purposes. No meaningful portion of any of the records could be segregated and released without identifying the communications underlying the communications intelligence report. Also disclosed would be the communications lines, channels, links, and systems targeted and intercepted and NSA's capabilities to successfully process the underlying communications. These communications targets and the processing techniques are current intelligence sources and methods. Disclosing them would permit foreign intelligence officials to draw inferences and make assessments about this nation's COMINT collection and processing activities that would enable them to take counter-measures, as described above, to defeat the capabilities of NSA's intelligence gathering techniques.

17. The COMINT reports being withheld from the plaintiff are classified in their entirety to protect intelligence sources and methods. When originated, certain of the records were properly classified Top Secret pursuant to Executive Order 10501, Section 1(a), providing for the application of that classification to information, the unauthorized disclosure of which could result in exceptionally grave damage to the nation, such as the "compromise of ... intelligence operations ... vital to the national defense." Other documents were properly classified Secret pursuant to Executive Order 10501, Section 1(b), providing for the application of that classification to information, the unauthorized disclosure of which could result in

serious damage to the nation, such as by "compromising... information revealing important intelligence operations." The rest, which were originated after June 1, 1972, were properly classified Top Secret pursuant to Executive Order 11652, Section 1(A), providing for the application of that classification to information, the disclosure of which could result in exceptionally grave damage to the national security, such as "the compromise of complex cryptologic and communications intelligence systems," or were properly classified Secret under Executive Order 11652, Section 1(B), providing for the application of that classification to information, the unauthorized disclosure of which could result in serious damage to the national security, such as the revelation of significant intelligence operations. Each COMINT record was appropriately marked when it was originated. Each COMINT record and each portion thereof remains properly so classified. These COMINT records meet the criteria for classification in Section 1-3 of Executive Order 12065, and they are properly classified within the categories provided in Section 1-1 of the Order. I have reviewed all of the COMINT records being withheld from the plaintiff for possible declassification or downgrading according to the provisions of Sections 3-1 and 3-3 of Executive Order 12065 and I find each portion of each record remains properly classified and therefore excluded from declassification or downgrading. In conducting this review, I weighed the significant need for openness in government against the likelihood of damage to our national security at this time and determined that the records should continue to be classified because of the damage their unauthorized disclosure would reasonably be expected to cause to communications intelligence activities of the United States Government. Because they are properly classified under Executive Order 12065, the records are exempt from disclosure pursuant to 5 U.S.C.§552(b)(1).

18. Release of the COMINT records being withheld from the plaintiff or any portion of any of them would disclose information about the nature of NSA's activities including its functions and thereby jeopardize the intelligence collection mission of the Agency. (See paragraphs 11 through 15 above.) This mission of the NSA is singular and unique. Public disclosure of specific information about the records in the context of that singular mission would reveal certain functions and activities of the NSA which are protected from disclosure by Section 6 of Public Law 86-36, 50 U.S.C. § 402 (note). Moreover, the disclosure of these classified records or of specific information about them would reveal information protected by 18 U.S.C. § 798 prohibiting the unauthorized disclosure of classified information concerning the communications intelligence activities of the United States. The disclosure of these records or any portion of them would also compromise classified information pertaining to intelligence sources and methods protected from disclosure by Section 103(d)(3) of the National Security Act of 1947 (50 U.S.C. § 403(d)(3)). Accordingly, it was determined that the COMINT records are exempt from release under 5 U.S.C. § 552(b)(3) of the Freedom of Information Act because each portion of each record is protected from disclosure by Section 6 of the Public Law 86-36, 18 U.S.C. § 798, and by Section 103(d)(3) of the National Security Act of 1947.

19. It should be noted that the classification of these COMINT records and the withholding of them pursuant to FOIA exemptions (b)(1) and (b)(3) are not based only on the substantive content of the documents but also on the characteristics of the reports that identify the intelligence sources and methods that would be seriously jeopardized by the disclosure plaintiff

seeks. That is, all of the COMINT reports at issue here (with one exception) are based on messages which were intercepted from government net communications systems. (See paragraph 10, above.) Moreover, almost all of these messages were enciphered when originally transmitted. Thus, release of any portion of the substantive message would not only risk identifying the ability of NSA to intercept a particular line of communications but would also risk revealing the capability of NSA to read a foreign government's enciphered messages. Similar harm would result from the disclosure of any material that might help to identify the communications intercepted by NSA, such as information about date, time, origin or manner of transmission or receipt. Also, the revelation of the substantive content of the reports would allow foreign officials to determine which channels or types of communications are being monitored. The public disclosure of either the content of the reports or of any identifying characteristics would have the same adverse consequences on the communications intelligence activities of the United States. All such information relates to classified communications intelligence functions of NSA that have not been publicly disclosed by the Agency in any other context.

RELEASE OF NON-COMINT RECORDS

20. Three of the four non-COMINT records at issue here were released in large part, with certain deletions. One of these records was withheld in its entirety. These documents are:

a. A document entitled UFO Hypothesis and Survival Questions which was prepared by an Agency employee. The entire report has been released except for the name and organization of the preparer of the document. This is not a COMINT report and contains no reference to SIGINT activities. It is a draft of a monograph that was located in an Agency file where it had been retained for historical reference purposes. The deletion was made pursuant to 5 U.S.C. § 552(b)(3) which provides that the FOIA does not apply to matters that are specifically exempted from disclosure by statute. The applicable statute in this instance is Section 6 of Public Law 86-36 which specifically exempts from disclosure the names and titles of NSA employees.

b. The second non-COMINT document is a memorandum which discusses the UFO phenomena as the author believes they may relate to the intelligence community. This document was released with deletions of the descriptive references to the COMINT operations of the Agency. The deletions contain information which I have determined to be currently and properly classified and, thus, exempt from disclosure under 5 U.S.C. § 552(b)(1). Portions of the material deleted also concern information with respect to the organization and operational activities and functions of NSA which are exempt from disclosure pursuant to 5 U.S.C. § 552(b)(3) which provides that the FOIA does not apply to matters that are specifically exempted from disclosure by another statute. Section 6 of Public Law 86-36 provides that no law shall be construed to require disclosure of the organization or any function of the NSA or any information with respect to the activities thereof. Moreover disclosure of this information would reveal information protected by 18 U.S.C. §798 which prohibits the unauthorized disclosure of classified information concerning COMINT activities and by the National Security Act of 1947 (50 U.S.C. §403(d)(3)) which prohibits the disclosure of information pertaining to intelligence sources and methods. A portion of this document was deleted pursuant to 5 U.S.C. §552(b)(5) because it represents this employee's expression of opinion on how the topic relates to the mission of the Agency. This deletion is non-factual and does not represent finalized Agency policy. It includes the kind of analysis, frank comment and recommendations, which an agency must encourage and protect from public disclosure to avoid a chilling effect upon free and candid internal discussions in support of optimum decision making within the Agency. Finally, a portion of this record in addition to being exempt from disclosure under 5 U.S.C. §552(b)(1) and (b)(3) is being withheld for the reason that it is not within the scope of plaintiff's request.

c. The third non-COMINT document is a memorandum for the record by an NSA assignee that was originally withheld in its entirety pursuant to 5 U.S.C. § 552(b)(5) and (b)(6). In my review today I have ascertained, however, that this memorandum is neither in whole nor in part responsive to the plaintiff's request. It does not deal with UFOs or the UFO phenomena. Rather, it is a document voluntarily prepared by the assignee to report an incident that occurred during his attendance at a UFO symposium. It is the assignee's personal account of his activities and does not include reference to any UFO sighting or phenomena.

d. The final non-COMINT record is a report which was addressed in paragraph 8, above. The portion of the record responsive to this FOIA request has been released to plaintiff's attorney, (See Exhibit 8, attached hereto). The remaining portions of the record contain no reference relating to UFOs or UFO phenomena and are, therefore, not responsive to plaintiff's request.

21. Further information about the records or portions thereof being withheld is contained in the in camera affidavit executed by me.

Eugene F. Yeates
EUGENE F. YEATES
Chief, Office of Policy
NSA

Subscribed and sworn to before me this 30th day of September 1980.

Gerald S. Everett
NOTARY PUBLIC

My commission expires on July 1, 1982.

14

Citizens Against Unidentified Flying Objects Secrecy v. NSA

(Civil Action No. 80-1562)

* In Camera Affidavit of
Eugene F. Yeates

TOP SECRET Supplemental Tesimony
Supporting Previous Public Testimony

This page intentionally left blank.

[redacted]

UNITED STATES DISTRICT COURT
FOR THE DISTRICT OF COLUMBIA

CITIZENS AGAINST UNIDENTIFIED FLYING OBJECTS SECRECY, Plaintiff, v. NATIONAL SECURITY AGENCY, Defendant.	)	Civil Action No. 80-1562

IN CAMERA
AFFIDAVIT OF EUGENE F. YEATES

County of Anne Arundel)
) ss:
State of Maryland)

Eugene F. Yeates, being duly sworn, deposes and says:

1. (U) I am the Chief, Office of Policy, of the National Security Agency (NSA). As Chief, Office of Policy, I am responsible for processing all initial requests made pursuant to the Freedom of Information Act (FOIA) for NSA records. The statements herein are based upon personal knowledge, upon my personal review of information available to me in my official capacity, and upon conclusions reached in accordance therewith.

2. (U) This affidavit supplements my unclassified affidavit executed on September 30, 1980 regarding all documents which have been located by NSA pursuant to plaintiff's FOIA request but which have been withheld wholly or in part by NSA. I submit this affidavit <u>in camera</u> for the purpose of stating facts, which cannot be publicly disclosed, that are the basis for exempting the records from release to the plaintiff.

3. ([redacted]CCO) At the beginning of each paragraph of this affidavit, the letter or letters within parentheses designate(s) the degree of sensitivity of information the paragraph contains.

[redacted]

The letters "U", "C", "S" and "TS" indicate respectively that the information is unclassified or is classified CONFIDENTIAL, SECRET or TOP SECRET. The symbols "(SC)" and "(TSC)" stand for "SECRET CODEWORD" and "TOP SECRET CODEWORD", respectively. "CODEWORD" refers to one of the distinctive five-letter words used to identify the source of the information as communications intelligence (COMINT), to distinguish between COMINT categories and sub-categories, and to facilitate the application of regulations for the dissemination and use of COMINT. [The codeword "UMBRA" appearing in conjunction with the TOP SECRET classification at the top and bottom of each page of this affidavit, is the codeword applicable to Category III (the highest category) COMINT.] Documents revealing sensitive details about the production of COMINT must bear the classification and codeword appropriate to the highest category or sub-category of COMINT to which they relate, even though they may not contain COMINT as such. [The symbol "CCO", which stands for the caveat "HANDLE VIA COMINT CHANNELS ONLY", is used to designate information related to COMINT or COMINT activities, which, although it does not require codeword protection, must be kept within COMINT channels, i.e., disclosed only to persons eligible to receive COMINT itself.]

THE RELEVANT DOCUMENTS

4. [redacted] In processing the plaintiff's FOIA request, a total of two hundred and thirty-nine documents were located in NSA files. Seventy-nine of these documents originated with other government agencies and have been referred by NSA to those agencies for their direct response to the plaintiff. One document, which I addressed in paragraph 20c of my public affidavit, was erroneously treated as part of the subject matter of plaintiff's FOIA request. It is an account by a person

assigned to NSA of his attendance at a UFO symposium and it cannot fairly be said to be a record of the kind sought by the plaintiff. Another document, discussed in paragraph 20d of my public affidavit, was recently declassified and released to plaintiff. Two additional non-COMINT records have been released to the plaintiff with the exempted material deleted. The deletions in these documents are explained below:

a. A document entitled <u>UFO Hypothesis and Survival Questions</u> was released to the plaintiff with the deletion on page seven of the name of the employee who prepared the draft and a deletion of a reference to his NSA component. As I explained in paragraph 20, sub-paragraph a, of my open affidavit, information about NSA's organization or employees is protected from disclosure by Public Law 86-36 and, therefore, exempt pursuant to 5 U.S.C. §552(b)(3).

b. The second non-COMINT document is a three page undated, unofficial draft of a monograph with a four page appendix by the same agency employee who authored the draft referenced in sub-paragraph a, above. This document was discussed in paragraph 20b of my public affidavit. It is entitled <u>UFO's [and the Intelligence Community Blind Spot to Surprise or Deceptive Data.]</u> In this document, the author discusses what he considers to be a serious shortcoming in the Agency's COMINT interception and reporting procedures -- [the inability to respond correctly to surprising information or deliberately deceptive data.] He uses the UFO phenomena to illustrate his belief that [the inability of the U.S. intelligence community to process this type of unusual data adversely affects U.S. intelligence gathering capabilities.] Deletions in this document were made as follows:

(1) All of the title after UFO, which addresses the perceived shortcoming, and all of paragraph one, which discusses [the employee's perception of the negative implications]

of the handling of UFO phenomena as it demonstrates what he believes is the [less than optimum ability of the intelligence community to process and evaluate highly unusual data.] As I stated in my public affidavit (paragraph 20b), the type of candor that is reflected in this record must be encouraged especially in an intelligence Agency where the most meaningful suggestions regarding ways to promote the efficiency of the critical Agency mission will of necessity come from within. Public disclosure of such information, especially when it advances a novel theory, could have the effect of stifling such candor by the risk of diminution of professional standing the employee runs if subsequently found wrong. Thus, this matter was deleted pursuant to 5 U.S.C. §552(b)(5).

(2) Paragraph three of this document uses a signals intelligence operation [against] [redacted] to illustrate the author's point. This paragraph contains information about SIGINT activities that is currently and properly classified and, thus, is exempt from disclosure pursuant to 5 U.S.C. §552(b)(1). The material in this paragraph also concerns the organization and operational activities and functions of NSA [directed against] [redacted] This material is exempt from disclosure under 5 U.S.C. §552(b)(3) which exempts from release under the FOIA matters specifically exempted from disclosure by another statute. As noted in paragraph 20, sub-paragraph b of my public affidavit, Public Law 86-36 provides that no law shall be construed to require disclosure of the organization or any function of the NSA or any information with respect to activities thereof.

(3) Paragraph four of the memorandum states the conclusions and recommendations of the author. While it talks of the ability of the Agency employees to deal with unusual phenomena it is not responsive to the plaintiff's request

regarding UFO or UFO phenomena. In any event, as I stated in my public affidavit (paragraph 20b), the subject matter of that paragraph is exempt from disclosure because it contains the employee's specific recommendations for addressing the problem of responding to surprise material. For the reasons stated in sub-paragraph (1) above, these recommendations are exempt from disclosure pursuant to 5 U.S.C. §552(b)(5). One specific recommendation suggests an operational approach to solving the problem which reveals NSA activities and is, therefore, exempt from disclosure pursuant to 5 U.S.C. §552(b)(3) as explained above.

(4) The final deletion is in appendix A, paragraph 10 of this report. This section talks [about deceptive communications tactics used by] the Vietnamese against U.S. forces and does not include any reference to UFO or UFO phenomena and is, therefore, not responsive to plaintiff's request. Nonetheless the subject matter of sub-paragraph 10 is currently and properly classified. Thus, even if it were deemed to be within the scope of plaintiff's request, it is exempt from disclosure pursuant to 5 U.S.C. §552 (b)(1).

COMINT REPORTS

5. (TSC) The remaining one hundred and fifty-six records being withheld are communications intelligence (COMINT) reports which were produced between 1958 and 1979. For purposes of my discussion here, these records are organized into three groups based upon the source of the report.

a. One hundred and fifteen of these reports were produced by [the signals intelligence organizations] [redacted] [These COMINT reports are provided to NSA] [redacted]

5

[redacted]

(1) Two of the records at issue here were produced by [redacted] [The United States has extremely] [redacted] [The report from] [redacted] [was provided to NSA] [redacted]

(2) One hundred and thirteen reports were provided [redacted]

b. Two of these COMINT reports originated from SIGINT operations which [redacted]

[redacted]

in exchange for the sharing of technology and COMINT information.

c. The remaining thirty-nine COMINT documents were produced by NSA or relate to NSA SIGINT operations. That is, these reports originated at NSA itself or in field sites under the operational and technical control of NSA.

6. (U) All of the COMINT reports are in either message or summary format. A report in message format contains a single underlying communication presented in a classic cable format, i.e., the verbatim text of the particular transmission, preceded and followed by "externals" consisting of: data about the sender and the recipient; the dates and times of transmission; and other technical information. A summary, as the label suggests, provides in summary form the contents of a single message or of a small number of related intercepted communications, often accompanied by some technical data.

7. [redacted] One hundred and fifty-four of the one hundred and fifty-six COMINT reports are based wholly upon intercepted communications of foreign governments transmitted on non-public "government net" communications links or systems. Of the two reports not included in this total, one report is the text of a [redacted] communication transmitted by an international communications common carrier. I have described the distinction between these two kinds of communications facilities in my public affidavit at paragraph 10. The other record which is not based on intercepted communications from "government net" facilities,

7

is a description of an incident purported to have been learned from U.S. interception and analysis of [redacted] communications.

8. ([redacted]) The COMINT reports originated by [redacted] can be further described in terms of sources and intelligence targets as follows:

a. [redacted] COMINT Reports Which Target Communications Transmitted on [redacted] Communications Facilities. One [redacted] report, in summary format, was produced by [redacted] from the intercepted communications between two [redacted] aircrafts and a ground controller in [redacted] 1971. [redacted] report a "phenomena" in the sky north of [redacted]. Three [redacted] reports are summaries of [redacted] messages intercepted from [redacted] communications. The messages were transmitted from [redacted] radar operators to a central control station. In these [redacted] messages, the [redacted] operators report everything that appears on their radar screens. When they cannot identify a particular object, they report it as an unidentifiable object. In translating these messages, the U.S. cryptolinguist uses "unidentified flying object" as the equivalent of the [redacted] text.

b. [redacted] COMINT Reports Which Target the [redacted] Communications System of [redacted]. One hundred and nine documents in summary format report on intercepted communications between [redacted] radar operators and a central control point. These summaries are similar to the reports described in sub-paragraph a above and,

8

[redacted]

again, contain [redacted] information pertaining to the radar operator's report of objects on his radar screen which he cannot identify.

c. [redacted] COMINT Report Which Targets [redacted] Communications. This document is a summary of intercepted messages prepared in 1976 which report radar tracking information from a [redacted] radar station to a central control point. It is similar to the reports described in sub-paragraphs (a) and (b) above.

d. [redacted] COMINT Report Which Targets [redacted] Communications. This summary was prepared in 1966 by [redacted] It contains a summary of intercepted communications which were transmitted between the [redacted] and the [redacted] commander regarding a yellow object that was reported to have fallen into the sea. These messages were transmitted along a [redacted] "government net" facility.

e. Two COMINT Reports Were the Product of [redacted] SIGINT Operations and Targeted the [redacted] Communications of [redacted]. The two reports were produced from a field site which is [redacted] operated by the U.S. and [redacted] The reports were prepared in 1966 and contain summaries of the communications transmitted by [redacted] radar operators as discussed in sub-paragraph b, above. One of the two reports is a follow-up report to the other. The relevant activity reported deals with the tracking of an object approximately 50 nautical miles northeast of [redacted] by the [redacted] Facility. As with the messages described above, these COMINT reports are summaries of the [redacted] messages.

[redacted]

~~TOP SECRET UMBRA~~

9. [redacted]SC) NSA-originated reports - Thirty-eight documents are the direct product of NSA SIGINT operations and one document describes classified SIGINT activities. These documents can be further described as follows:

a. The document describing SIGINT operations reports an alleged intercept of [redacted] communications. The factual circumstances of the incident reported in this record were received by NSA from an FOIA requester other than the plaintiff and are considered to be fictitious by NSA analysts. [redacted]

b. One record is a 1973 report which summarizes the [redacted] text of [redacted] message transmitted [redacted] from the [redacted] to the [redacted] in [redacted] It makes reference to a purported UFO sighting by [redacted]

10

[redacted]

c. Twelve NSA-originated COMINT reports target the communications links and systems of [redacted] Two documents, in summary format, report the [redacted] communications. Two of the records are in [message] format and report the [redacted] communications which relate that an unidentified flying object was sighted in the air by a [redacted] unit. One report contains a summary of [redacted] activity based upon communications in reaction to an unidentified flying object along the [redacted] Two documents report on communications transmitted between [redacted] and [redacted] who report visual observations of luminous spheres. One report is a summary of a transmission between [redacted] Finally, four documents in this group of twelve were intercepted from other [redacted] communications targets. One document is based on the intercepted transmission of a [redacted] reporting a bright light. The second record is based on the intercept of a transmission of an [redacted] to a [redacted] station seeking a report on any shining phenomena or falling meteorites observed on specified dates. The third and fourth reports are a summaries of on-going debates on UFOs among [redacted] based on intercepted communications transmitted on [redacted] nets.

d. Five of the NSA-originated COMINT reports target [government net] communications [redacted] All five of these documents are based on intercepted [redacted] communications [redacted] units and [redacted] reporting observations of luminous objects in the sky.

e. Four NSA-originated COMINT reports target [redacted] communications links or systems. Three reports are in summary format and are based on an intercepted message transmitted [redacted] sites reporting an unidentified flying object at a very high altitude; an intercepted message transmitted from [redacted] reporting that an unidentified flying object with two lights had passed over [redacted] and an intercepted voice message transmitted to [redacted] facility by a [redacted] who reported an unidentified flying object. The fourth report is based upon a message between [redacted] units regarding a UFO sighting.

f. Sixteen NSA-originated COMINT reports target the "government net" communications systems and links of several different countries. This group contains summaries of intercepted transmissions between [redacted] (two reports based on communications of [redacted] and an unidentified sender reporting sightings of a UFO), [redacted] (two reports based on communications by [redacted] reporting unidentified flying objects), [redacted] (a report based on a message from a [redacted] to an unidentified receiver in [redacted] reporting a sighting of unidentified flying objects), [redacted] (a report from an [redacted] to the [redacted] in [redacted] reporting an object that appeared to be a rocket over [redacted] (a report based on communications between two [redacted] and an unidentified [redacted] station in which [redacted] reported sightings of bright or light spots), [redacted] (a report based on a transmission from a [redacted] unit to a [redacted] reporting a sighting of an unidentified flying object), [redacted] (a report based on a message from [redacted] to an an unidentified recipient which provided instructions for reporting the sighting of flying objects), and [redacted] (a report based upon a transmission between [redacted] reporting that some [redacted] saw a ball of light about the size of an orange moving overhead). One document in summary format is the product of an intercepted [redacted] transmission reporting the sighting of an elongated ball of fire. One document in message format reports the text of a [redacted] message sent by the [redacted] to the [redacted] reporting an unidentified flying object. One document in summary format reports the [redacted] text of [redacted] message from [redacted] to [redacted] which was transmitted [along a "government net"] facility. It reports, among other items, an increase in UFO activity. The last three documents in this group report on intercepted communications of [redacted] Each report contains information derived from intercepted transmissions reporting the tracking of unidentified aircraft by [redacted] [radar] (STET) operators.

EXEMPTION OF THE COMINT REPORTS

10. [redacted] A primary and often overriding consideration regarding the classification of COMINT reports is that the need to protect communications intelligence sources and methods is greater than the need to protect sensitive contents of the underlying intercepted messages. Nevertheless, no portion of the contents of COMINT reports may be disclosed, where, as here, revealing the information would have the effect of identifying

for the target communicators the specific communications that had been intercepted and exploited. One hundred and fifty-four of the COMINT reports being withheld are the product of intercept operations directed against foreign government controlled communications systems within their territorial boundaries. Revealing the contents of these reports would disclose the capability of NSA to target these government controlled communication systems. Even where the underlying communications are not specified, [redacted] the communications had been intercepted for processing by NSA. Moreover, the disclosure of these reports would reveal much more than the identity of the targeted communications systems. It would reveal as well [redacted]

11. (TS-CCO) The communications sources involved in this case -- which are specified or implicitly identified in the COMINT reports being withheld by NSA -- are the source of extremely valuable communications intelligence covering a broad range of kinds of information from [redacted] and other [redacted] activities to [redacted] and [redacted] matters. Release of these documents would seriously damage the ability of the United States to gather this vital intelligence information for the following reasons:

a) Disclosure of the report discussed in paragraph 9b would inform [redacted] that their communications by international common carrier facilities [redacted] can be intercepted and selected out by NSA, and, even more importantly, that the [redacted]

[redacted] Furthermore, revealing the NSA intercept operation against the international common access carrier route which yielded the [redacted] message could cause the United States to lose access to the communications of other foreign governments who use that same communications route.

(b) The disclosure of reports, such as [redacted] one at issue here, based on messages transmitted via international common access carrier facilities would also reveal this Agency's capacity to select from such intercepted communications those messages having potential intelligence value. [redacted] NSA usually [redacted]

(c) As I have stated in my open affidavit, when alerted to the extent of NSA's capability, and if given information from

[redacted]

which inferences could be drawn as to the processing methods used, foreign intelligence services would be able to evade or defeat portions of NSA's present foreign intelligence efforts [targeting] international common carrier links. These countries could be expected to use different routes of communication or [redacted] [The costs involved would be substantial but not prohibitive; the technology required is now available.] [redacted]

12. The disclosure of other records at issue here, would result in the loss of the intelligence information gathered from the interception of the government net communications systems. The value of the intelligence data collected from these sources is obvious.

(a) [For example, analysis of data collected from the intercept of] [redacted] [communications] [redacted] [of a foreign government—] [redacted] (paragraphs 8a, b, c, e and 9f) [redacted] communications (paragraph 9c), or transmissions between [redacted] (paragraphs

16

[redacted]

8d and 9d, e, f) -- immeasurably aids U.S. analysts' studies of the [redacted] Analysts are able to report on the [redacted] capabilities of foreign [redacted] generally. This information enables planners in turn to assess the capability of [redacted] system [redacted] and [redacted] The data transmitted [redacted] communications is useful in evaluating the performance capabilities of [redacted] By monitoring the [redacted] activity and other [redacted] transmissions and relating it to geographic areas, the U.S. analyst can [redacted] Other targeted [redacted] communications provide critical technical information, such as data [redacted] which is vital to the development of U.S. countermeasures. Foreign [redacted] communications are among the [redacted] sources of intelligence information regarding their nation's intentions, in both the short and long terms. Moreover, the government net systems described here continue to yield valuable intelligence data -- including the [redacted] communication systems discussed in paragraph 9d. [redacted]

(2) Also, the data collected from intercept operations against [redacted] government net systems (described in paragraph 9c) provides invaluable information to our policymakers. From these sources U.S. analysts are able to compile reports on significant [redacted] activity, [redacted] and [redacted] events. Also, by monitoring these [redacted] transmissions, analysts are able to obtain data to

[redacted]

[redacted] The intelligence collected from these sources is no less vital to U.S. planners than the [redacted]

13. (TSC) The need to protect against any identification of the targets of intercept operations is equaled by the need to protect against revealing the identity of the [redacted] sources [redacted] Disclosure of the records [redacted] could have extremely adverse repercussions to the U.S. over and beyond the [redacted] The most serious [redacted] would be a [redacted] At the very least [redacted] significant loss of [redacted] Furthermore, disclosures which would tend to identify [redacted] for COMINT [redacted]

[redacted]

[redacted]

CLASSIFICATION OF THE WITHHELD COMINT REPORTS

14. (S-[redacted]) As I have indicated in paragraph 17 of my open affidavit, I have determined that the one hundred and fifty-six [redacted] reports relating to COMINT activities at issue here are based on intercepted communications of foreign governments or SIGINT operations and, thus, remain properly classified. In conducting this review I have weighed the significant need for openness in government against the likelihood of damage to our national security at this time and have determined that each record should continue to be classified. No meaningful portion can be segregated from the records without revealing classified information about the intercepted communications underlying the COMINT reports. Because each record and each portion thereof is properly classified under Executive Order 12065, it is exempt from disclosure pursuant to 5 U.S.C. §552(b)(1).

15. (S-[redacted]) The interception, processing and exploiting of foreign communications sent on international common carrier facilities or by government net channels are within the COMINT mission of NSA. So, too, is the carrying out of [redacted] These functions and activities of NSA are particular types of matters that may be withheld under 5 U.S.C. §552(b)(3), since Section 6

[redacted]

of Public Law 86-36 permits the Agency to refuse to release them. In this case, the COMINT reports reflecting those functions and activities must be withheld to avoid compromising the efficacy of the sources of COMINT information involved.

16. [redacted] Information about the interception, processing and exploitation of the foreign communications underlying the records being withheld by NSA is classified information concerning communications intelligence activities of the United States [redacted] the unauthorized disclosure of which is prohibited by 18 U.S.C. §798, paragraphs (a)(3) and (a)(4). This information because it is prohibited from disclosure by statute, is exempt from release under the FOIA pursuant to 5 U.S.C. §552(b)(3).

17. (U) As stated in my public affidavit, the information that would be disclosed by these records is information about intelligence sources and methods protected from unauthorized disclosure under 50 U.S.C. §403(d)(3). The reports are therefore exempt from release under Exemption 3 of the FOIA. 5 U.S.C. §552(b)(3).

18. (U) In view of the foregoing, and in order to protect existing sensitive and important foreign intelligence sources and processing techniques vital to the national security, I certify that disclosure of past and present foreign intelligence communications activities of NSA revealed in the records the plaintiff seeks would endanger highly valuable sources of foreign intelligence.

19. (U) Finally, I respectfully request that the Court treat this affidavit in the same secure manner as it has been handled in submission to the Court, and to return it to appropriate personnel of the Department of Justice as soon as possible after review by the Court. The Department

[redacted]

of Justice will retain custody of this document under the Court's seal, subject to any further orders of this Court or any other court of competent jurisdiction.

Eugene F. Yeates
EUGENE F. YEATES
Chief, Office of Policy

Subscribed and sworn to before me this
9th day of October 1980.

NOTARY PUBLIC

My commission expires on July 1, 1982.

21

[redacted]

This page intentionally left blank.

* Communications Intelligence (COMINT) Reports

TOP SECRET Documents
Redacted and Released
by the NSA as a Result of
Civil Action No. 80-1562

This page intentionally left blank.

>>>>>>>>>>>>>>>>>>>>>>>>>> ~~TOP SECRET CODE~~[redacted] <<<<<<<<<<<<<<<<<<<<<<<<<
*** BEGIN MESSAGE 49 ***
SERIAL=A14/M1021/79073/80505 UDN=Q01MAR79-30APR79(11552)
DE [redacted] #0009 0730717
ZNY MMNSH
ZKZK PP [redacted] DE
[redacted]
ZEM
~~S E C R E T S P O K E~~ [redacted]
~~WNINTEL~~ SECTION SEVEN
[redacted]
THIS MESSAGE IS CLASSIFIED SECRET SPOKE IN ITS ENTIRETY.
THE FOLLOWING INFORMATION HAS BEEN PROVIDED [redacted]
[redacted]
1355-1401, ONE UFO MOVED FROM 75 NAUTICAL MILES EAST
OF [redacted] TOWARD EAST AND FADED NEAR [redacted]

1

>>>>>>>>>>>>>>>>>>>>>>>>> ~~TOP SECRET CODEWORD~~ <<<<<<<<<<<<<<<<<<<<<<<<<<<

TOP SECRET [redacted]

CATEGORY = 400 M = 10

MESSAGE = 03787241

LH R0011 0540631
ZNY MMNSH
ZKZK PP [redacted] DE
[redacted]
ZEM
[redacted] UNINTEL SECTION 4 [redacted] THIS
MESSAGE IS CLASSIFIED SECRET SPOKE IN ITS ENTIRETY. THE FOLLOWING INFORMATION
HAS BEEN PROVIDED [redacted]

[redacted]

2

[redacted]

.. RADAR TRACKING OF
UNIDENTIFIED FLYING OBJECTS (UFO): A. Ø93Ø-Ø933, ONE UFO (PROBABLY A BALLOON)
MOVED FROM 22 NM SE [redacted] TOWARD EAST, ALT 49,2ØØ FT. REVW [redacted]
[redacted] #ØØ11 NNN

[redacted]

3

[redacted]

CATEGORY = 4ØØ M- = 9

MESSAGE = Ø3787232

LH #ØØØ6 Ø53Ø651
ZNY MMNSH
ZKZK PP [redacted] DE
[redacted]
ZEM
[redacted] SECTION 6 [redacted] THIS
MESSAGE IS CLASSIFIED SECRET SPOKE IN ITS ENTIRETY [redacted]
HAS BEEN PROVIDED [redacted]

6. RADAR TRACKING OF UNIDENIFIIED FLYING OBJECT (UFO): Ø6Ø1-1Ø4Ø, ONE UFO (PROBABLY A BALLOON) MOVED FROM [redacted] SOUTH OF [redacted] TOWARD EAST AND PASSED NORTH OF [redacted] ALT 46,ØØØ- 49,2ØØ FT.

[redacted]

4

[redacted]

CATEGORY = 4ØØ M = 7

MESSAGE = Ø3769694

LH #ØØ1Ø Ø43Ø841
ZNY MMNSH
ZKZK PP [redacted] DE
[redacted]
ZEM
[redacted] ~~WNINTEL~~ SECTION 6 [redacted] THIS
MESSAGE IS CLASSIFIED SECRET SPOKE IN ITS ENTIRETY. ~~THE FOLLOWING INFORMATION~~
HAS BEEN PROVIDED [redacted]

[redacted]

5

[redacted]

7. RADAR TRACKING OF UNIDENTIFIED FLYING OBJECTS (UFO): A. Ø935-1Ø26, ONE UFO (PROBABLY A BALLOON) MOVED FROM [redacted] TOWARD NE AND FADED [redacted] ALT 46,ØØØ-59,2ØØ FT. B. 1317-1935, TWO UFO (PROBABLY BALLONS) MOVED FROM [redacted] AND [redacted] RESPECTIVELY TOWARD NE, PASSED [redacted] SOUTH OF [redacted] AND SOUTH OF [redacted] AND THEN FADED, ALT 44,ØØØ-49,2ØØ FT [redacted]

TREVW [redacted]

[redacted] #ØØ1Ø NNN

[redacted]

[redacted]

CATEGORY = 4ØØ M = 98

MESSAGE = Ø3749175

LII #ØØ13 361Ø7Ø3
ZNY MMNSH
ZKZK PP [redacted] DE
[redacted]
ZEM
[redacted] SECTION FIVE [redacted] THIS
MESSAGE IS CLASSIFIED SECRET SPOKE IN ITS ENTIRETY. THE FOLLOWING INFORMATION
HAS BEEN PROVIDED [redacted]

6. RADAR TRACKING OF
UNIDENTIFIED FLYING OBJECTS (UFO): 1Ø41-1Ø51, TWO UFO (PROBABLY BALLOONS)

[redacted]

7

[redacted]

MOVED 43 NAUTICAL MILES 19Ø DEGREES OFF [redacted] AND FADED 22 NAUTICAL MILES
17Ø DEGREES OFF [redacted]

REVW [redacted]. #ØØ13 NNN

[redacted]

8

[redacted]

CATEGORY = 4ØØ M = 96

MESSAGE = Ø3749164

LH #ØØØ9 36ØØ619
ZNY MMNSH
ZKZK PP [redacted] DE
[redacted]
ZEM
S E C R E T SPOKE [redacted] UNINTEL SECTION 6 [redacted] THIS MESSAGE IS CLASSIFIED SECRET SPOKE IN ITS ENTIRETY. THE FOLLOWING INFORMATION HAS BEEN PROVIDED [redacted]

6. RADAR TRACKING OF UNIDENTIFIED FLYING OBJECT (UFO): Ø95Ø-1349, ONE UFO (PROBABLY A BALLOON) MOVED FROM [redacted], PASSED [redacted] AND [redacted] AND FADED 19Ø NAUTICAL MILES EAST OF [redacted] ALT 52,5ØØ FT.

[redacted]

9

~~T O P S E C R E T U M B R A~~

CATEGORY = 400 M = 95

MESSAGE = 03749154

LH #0014 3560711
ZNY MMNSH
ZKZK PP [redacted] DE
[redacted]
ZEM
~~S E C R E T SPOKE~~ [redacted] WNINTEL SECTION SEVEN [redacted] THIS MESSAGE IS CLASSIFIED SECRET SPOKE IN ITS ENTIRETY. THE FOLLOWING INFORMATION HAS BEEN PROVIDED [redacted]

6. RADAR TRACKING OF UNIDENTIFIED FLYING OBJECT (UFO): 0854-0930, ONE UFO (PROBABLY A BALLOON) MOVED FROM 32 NM 90 DEG OFF [redacted] TOWARD EAST AND FADED 64 NM 90 DEG OFF [redacted], ALT 47,640-49,200 FT.

[redacted]

10

[redacted]

CATEGORY = 4ØØ M = 94

MESSAGE = Ø37491Ø7

LH #ØØØ9 353Ø413
ZNY MMNSH
ZKZK PP [redacted] DE
[redacted]
ZEM
S E C R E T SPOKE [redacted]NTEL SECTION 6 [redacted] THIS MESSAGE IS CLASSIFIED SECRET SPOKE IN ITS ENTIRETY. THE FOLLOWING INFORMATION HAS BEEN PROVIDED [redacted]

7.
RADAR TRACKING OF UNIDENTIFIED FLYING OBJECTS (UFO): Ø9ØØ-1ØØØ, 13 UFO MOVED FROM [redacted], PASSED [redacted] AND FADED [redacted]

TOP SECRET UMBRA

11

TOP SECRET UMBRA

CATEGORY = 4ØØ M = 89

MESSAGE = Ø37ØØ643

LH #ØØØ8 279Ø65Ø
ZNY MMNSH
ZKZK PP [redacted] DE
[redacted]
ZEM
S E C R E T SPOKE [redacted] UNINTEL SEC 8 [redacted] THE FOLLOWING
INFORMATION HAS BEEN PROVIDED [redacted]

TOP SECRET UMBRA

12

~~TOP SECRET UMBRA~~

5. RADAR TRACKING OF UNIDENTIFIED FLYING OBJECTS(UFO): 1420-1600, THREE UFO (PROBABLY BALLOONS) MOVED FROM [redacted] TOWARD WEST SLOWLY AND FADED [redacted] ALT 69,000-79,000 FT.

[redacted]

13

[redacted]

CATEGORY = 4ØØ M = 88

MESSAGE = Ø37ØØ623

LII #ØØØ7 277Ø639
ZNY MMNSH
ZKZK PP [redacted] DE
[redacted]

ZEM
S [redacted] FINAL SECTION OF 3 [redacted] THE
FOLLOWING INFORMATION HAS BEEN PROVIDED [redacted]

6. RADAR TRACKING OF UNIDENTIFIED FLYING OBJECTS (UFO): A. Ø435-Ø52Ø, FOUR UFO (PROBABLY BALLOONS) MOVED FROM WEST OF [redacted] TOWARD SOUTH, PASSED EAST OF [redacted] AND FADED EAST OF [redacted] ALT 11,5ØØ-13,1ØØ FT. B. Ø455-Ø52Ø, ONE UFO (PROBABLY A BALLOON) MOVED FROM EAST OF [redacted] TOWARD SOUTH AND FADED EAST OF [redacted] ALT 11,5ØØ-13,1ØØ FT. C. Ø734-1Ø25, THREE UFO (PROBABLY BALLOONS) MOVED SLOWLY FROM SOUTH OF [redacted] TOWARD WEST AND PASSED [redacted] AND [redacted] ALT 49,2ØØ-66,ØØØ FT.

T [redacted]

14

[redacted]

CATEGORY = 4ØØ   M = 86

MESSAGE = Ø3689999

LH #ØØØ9 272Ø646
ZNY MMNSH
ZKZK PP [redacted] DE
[redacted]
ZEM
[redacted] " UNINTEL SEC 9 [redacted] THE FOLLOWING
INFORMATION HAS BEEN PROVIDED [redacted]

5. RADAR
TRACKING OF UNIDENTIFIED FLYING OBJECTS (UFO): Ø9Ø9-Ø93Ø, TWO UFO (PROBABLY
BALLOONS) MOVED FROM 43 NAUTICAL MILES SE OF [redacted] TOWARD WEST
AND FADED NEAR [redacted] ALT 54,ØØØ-1ØØ,ØØØ FT.

~~TOP SECRET UMBRA~~

15

[redacted]

CATEGORY = 4ØØ M = 85

MESSAGE = Ø3689955

LH #ØØ19 265Ø8Ø5
ZNY MMNSH
ZKZK PP [redacted] DE

[redacted]

ZEM
[redacted]E [redacted] " [redacted] SECTION 9 [redacted] THE
FOLLOWING INFORMATION HAS BEEN PROVIDED [redacted]

[redacted]

16

11

~~TOP SECRET UMBRA~~

7. RADAR TRACKING OF UNIDENTIFIED FLYING OBJECTS (UFO): A. 0900-1210, FOUR UFO (PROBABLY BALLOONS) MOVED SLOWLY FROM NW OF [redacted] TOWARD WEST, PASSED [redacted] AND ENTERED INTO [redacted] ALT 85,300 FT. B. 1400-1425, ONE UFO (PROBABLY A BALLOON) MOVED FROM [redacted] TOWARD WEST, ALT 15,000 FT. C. 1440-1451, TWO UFO (PROBABLY BALLOONS) MOVED OVER [redacted] ALT 11,500-15,000 FT. D. 0805-1450, FIVE UFO (PROBABLY BALLOONS) MOVED SLOWLY FROM [redacted] TOWARD WEST, ALT 72,100-91,800 FT.

[redacted]

17

T[redacted]

CATEGORY = 4ØØ M = 84

MESSAGE = Ø3689948

LH #ØØ11 264Ø754
ZNY MMNSH
ZKZK PP [redacted] DE
[redacted]
ZEM
[redacted] SECTION EIGHT [redacted]THE
FOLLOWING INFORMATION HAS BEEN PROVIDED[redacted]

T O P S E C R E T U M B R [redacted]

18

6. RADAR TRACKING OF UNIDENTIFIED FLYING OBJECTS (UFO): 0225-0946, 18 UFO (PROBABLY BALLOONS) MOVED FROM THE AREA BETWEEN [redacted] AND [redacted] TOWARD WEST, PASSED [redacted] AND ENTERED SOUTH OF [redacted] ALT 20,000- 69,000 FT

~~TOP SECRET UMB~~

19

[redacted]

ATEGORY = 4ØØ M = 83

MESSAGE = Ø3689914

UU #ØØØ3 262Ø642
ZNY MMNSH
ZKZK PP [redacted] DE

[redacted]

ZEM
[redacted] SECTION SIX [redacted] THE
FOLLOWING INFORMATION HAS BEEN PROVIDED [redacted]

[redacted] T UMBRA

20

[redacted]

6. RADAR TRACKING OF [UNIDENTIFIED FLYING OBJECTS (UFO): 0810-1329, 21 FUO (PROBABLY BALLOONS) AND 1620-1735, SEVEN UFO (PROBABLY BALLOONS) MOVED FROM SOUTH OF [redacted] AND PASSED [redacted] AND EAST OF [redacted] ALT 10,000-28,000 FT.]

[redacted]

21

[redacted]

CATEGORY = 4ØØ M = 81

MESSAGE = Ø3679Ø51

LH #ØØ14 258Ø655
ZNY MMNSH
ZKZK PP [redacted] DE
[redacted]
ZEM
S E C R E T SPOKE [redacted] UNINTEL SECTION TEN [redacted] THE
FOLLOWING INFORMATION HAS BEEN PROVIDED [redacted]

5. RADAR TRACKING OF UNIDENTIFIED FLYING OBJECTS (UFO): Ø95Ø-121Ø, EIGHT UFO (PROBABLY BALLOONS) MOVED FROM THE THE AREA NEAR [redacted] TOWARD WEST AND FADED [redacted] AND [redacted] ALT 72,1ØØ-88,58Ø FT. 6. [redacted] TO UNIDENTIFIED FLYING OBJECTS (UFO): Ø81Ø-Ø91Ø. 25 UFO (PROBABLY BALLOONS) MOVED FROM AREA BETWEEN [redacted] AND [redacted] TOWARD WEST AND PASSED [redacted] AND [redacted], ALT 16,5ØØ-85,3ØØ FT.

[redacted]

22

[redacted]

CATEGORY = 4ØØ   M = 8Ø

MESSAGE = Ø3679Ø34

LH #ØØ9Ø 257Ø7Ø1
ZNY MMNSH
ZKZK PP [redacted] DE
[redacted]
ZEM
[redacted] SECTION NINE [redacted] THE
FOLLOWING INFORMATION HAS BEEN PROVIDED [redacted]

[redacted]

23

[redacted]

26. RADAR TRACKING OF UNIDENTIFIED FLYING OBJECTS (UFO): A. Ø255-Ø746, SIX UFO (PROBABLY BALLOONS) MOVED SLOWLY FROM SW OF [redacted] TOWARD WEST AND PASSED [redacted] AND [redacted] ALT 66,ØØØ FT. B. Ø3ØØ-Ø62Ø; TWO UFO (PROBABLY BALLOONS) MOVED SLOWLY FROM NW OF [redacted] TOWARD WEST AND PASSED SOUTH OF [redacted] AND [redacted], ALT 56,ØØØ-69,ØØØ FT. C. Ø6Ø6-Ø6Ø6, FOUR UFO (PROBABLY BALLOONS) MOVED FROM [redacted] AND FADED SHORTLY AFTER, ALT 13,1ØØ FT. E2, IMPDET. #ØØ9Ø NNN

[redacted]

24

[redacted]

CATEGORY = 400 M = 79

MESSAGE = 03679008

LH #0015 2550651
ZNY MMNSH
ZKZK PP [redacted] DE
[redacted]
ZEM
[redacted] SECTION EIGHT [redacted] THE
FOLLOWING INFORMATION HAS BEEN PROVIDED [redacted]

8. RADAR TRACKING OF UNIDENTIFIED FLYING OBJECTS (UFO): A. 035-0430, ONE

[redacted]

25

T O P [redacted]

UFO (PROBABLY A BALLOON) MOVED FROM NW OF [redacted] TOWARD WEST AND FACED SOUTH OF [redacted] ALT 32,8ØØ-721,ØØØ FT. B. Ø54Ø-Ø81Ø, THREE UFO (PROBABLY BALLOONS) MOVED SLOWLY FROM SW OF [redacted] TOWARD WEST AND FADED SOUTH OF [redacted] ALT 56,ØØØ-66,ØØØ FT. C. Ø95Ø-17ØØ, 18 UFO (PROBABLY BALLOONS) MOVED SLOWLY FROM [redacted] TOWARD WEST AND FADED [redacted] ALT 69,ØØØ-72,1ØØ FT.

NNN

[redacted]

26

[redacted]

CATEGORY = 4ØØ M = 78

MESSAGE = Ø3678994

LH #ØØ19 254Ø712
ZNY MMNSH
ZKZK RP [redacted] DE
[redacted]
ZEM
[redacted] WN[redacted]EL SECTION SIX [redacted] THE
FOLLOWING INFORMATION HAS BEEN PROVIDED [redacted]

7. [redacted] TO

[redacted]

27

[redacted]

UNIDENTIFIED FLYING OBJECTS (UFO): ØØØ5-12Ø5; 72 UFO (PROBABLY BALLOONS) MOVED FROM WEST OF [redacted] TOWARD WEST AND PASSED [redacted] AND EAST OF [redacted] ALT 8,2ØØ- 75,5ØØ FT

[redacted]

28

[redacted]

CATEGORY = 4ØØ M = 77

MESSAGE = Ø3678974

LH #ØØ11 251Ø71Ø
ZNY MMNSH
ZKZK PP [redacted] DE
[redacted]

ZEM
S E C R E T SPO [redacted] UMINTEL SECTION FIVE [redacted] THE FOLLOWING
INFORMATION HAS BEEN PROVIDED [redacted]

[redacted]

29

[redacted]

6. RADAR TRACKING OF UNIDENTIFIED FLYING OBJECTS (UFO): A. ØØ28-Ø325, FOUR UFO (PROBABLY BALLOONS) MOVED SLOWLY FROM SE OF [redacted] TOWARD SW AND PASSED [redacted] AND [redacted]. B. Ø325-Ø515, ONE UFO (PROBABLY A BALLOON) MOVED SLOWLY FROM [redacted] TOWARD WEST, PASSED [redacted] AND FADED [redacted] C. Ø325-Ø515, ONE UFO (PROBABLY A BALLOON) MOVED SLOWLY FROM NORTH OF [redacted] TOWARD WEST AND FADED [redacted] D. 1355-163Ø, 19 UFO (PROBABLY BALLOONS) MOVED FROM [redacted] TOWARD WEST AND FADED [redacted] AND [redacted], ALT 69,ØØØ-79,ØØØ FT. E2, IMPDET. #ØØ11 NNN

TOP [redacted] UMBRA

30

T [redacted]

CATEGORY = 4ØØ M = 76

MESSAGE = Ø3678967

LH #ØØ14 25ØØ739
ZNY MMNSH
ZKZK PP [redacted] DE
[redacted]

ZEM /
S E [redacted] E [redacted] NINTEL SECTION SIX [redacted] THE
FOLLOWING INFORMATION HAS BEEN PROVIDED [redacted]

5.
RADAR TRACKING OF UNIDENTIFIED FLYING OBJECTS (UFO: A. 1Ø25-15Ø5, 26 UFO (PROBABLY BALLOONS) MOVED OVER [redacted] AND [redacted] ALT 69,ØØØ-98,6ØØ FT. B. 152Ø-1548, ONE UFO (PROBABLY A BALLOON) MOVED FROM THE AREA BETWEEN [redacted] AND [redacted] TOWARD WEST, AND PASSED [redacted] ALT 6,5ØØ FT. 6. [redacted] UNIDENTIFIED FLYING OBJECTS (UFO: Ø255-Ø92Ø, 21 UFO (PROBABLY BALLOONS) MOVED FROM THE AREA BETWEEN [redacted] AND [redacted] TOWARD WEST, AND PASSED [redacted] ALT 16,5ØØ-82,ØØØ FT.]

T [redacted]

31

[redacted]

CATEGORY = 4ØØ M = 75

MESSAGE = Ø3678942

LH #ØØ15 248Ø717
ZNY MMNSH
ZKZK PP [redacted] DE
[redacted]
ZEM
[redacted] SECTION EIGHT [redacted] THE
FOLLOWING INFORMATION HAS BEEN PROVIDED [redacted]

6. [redacted] UNIDENTIFIED FLYING OBJECTS (UFO):
Ø135-124Ø, 15 UFO (PROBABLY BALLOONS) MOVED FROM THE AREA BETWEEN [redacted] AND [redacted] TOWARD WEST AND PASSED [redacted] AND EAST OF [redacted] ALT 62,4ØØ-79,ØØØ FT.

[redacted]

32

[redacted]

CATEGORY = 4ØØ M = 72

MESSAGE = Ø3668993

LH #ØØ15 244Ø655
ZNY MMNSH
ZKZK PP [redacted] DE
[redacted]
ZEM
[redacted] SECTION NINE [redacted] THE
FOLLOWING INFORMATION HAS BEEN PROVIDED [redacted]

T [redacted]

33

TOP SECRET UMBRA

5. [redacted] UNIDENTIFIED FLYING OBJECTS (UFO):
ØØ14-Ø9ØØ, [redacted] RADAR TRACKED 23 UFO (PROBABLY BALLOONS) MOVING FROM THE AREA BETWEEN [redacted] AND [redacted] TOWARD WEST AND PASSING [redacted], E2, IMPDET. #ØØ15 NNN

T[redacted] SECRET UM[redacted]

34

[redacted]

CATEGORY = 4ØØ M = 73

MESSAGE = Ø3668994

LII #ØØ16 244Ø655
ZNY MMNSH
ZKZK PP [redacted] DE

[redacted]

ZEM
[redacted] SECTION TEN [redacted] THE
FOLLOWING INFORMATION HAS BEEN PROVIDED [redacted]

6. RADAR TRACKING OF UNIDENTIFIED FLYING OBJECTS (UFO): A.
Ø715-Ø811, ONE UFO(PROBABLY A BALLOON) MOVED OVER [redacted] AND
[redacted], ALT 7Ø,ØØØ FT. B. Ø856-Ø917, ONE UFO (PROBABLY A
BALLOON) MOVED OVER [redacted], ALT 87,6ØØ FT. 7.

[redacted]

35

~~TOP SECRET UMBRA~~

CATEGORY = 4ØØ M = 69

MESSAGE = Ø3668941

LH #ØØ11 24ØØ524
ZNY MMNSH
ZKZK PP [redacted] DE
[redacted]
ZEM
[redacted] (UNINTEL SECTION 4 [redacted] THE FOLLOWING INFORMATION HAS BEEN PROVIDED [redacted]

5. RADAR TRACKING OF UNIDENTIFIED FLYING OBJECTS (UFO): A. Ø458-Ø945, NINE UFO (PROBABLY BALLOONS) MOVED FROM NORTH OF [redacted] TOWARD WEST AND PASSED NE OF [redacted] AND [redacted] ALT 85,3ØØ-88,ØØØ FT. B. 1ØØ4-1118, FIVE UFO (PROBABLY BALLOONS) MOVED FROM [redacted] TOWARD WEST AND FADED [redacted] ALT 86,ØØØ FT. C. 1Ø5Ø-12Ø5, TWO UFO (PROBABLY BALLOONS) MOVED FROM [redacted] TOWARD WEST AND FADED NEAR [redacted] ALT 66,ØØØ FT.

~~TOP SECRET UMBRA~~

36

[redacted]

CATEGORY = 4ØØ   M = 63

MESSAGE = Ø3663928

LH #ØØ12 237Ø629
ZNY MMNSH
ZKZK PP [redacted] DE
[redacted]
ZEM
[redacted] E [redacted] UNINTEL SECTION NINE [redacted] THE
FOLLOWING INFORMATION HAS BEEN PROVIDED [redacted]

6. [redacted]
UNIDENTIFIED FLYING OBJECTS (UFO): A. Ø535-Ø71Ø, ONE UFO (PROBABLY A BALLOON) MOVED FROM [redacted] TOWARD WEST SLOWLY AND PASSED [redacted] AND [redacted] B. Ø54Ø-112Ø, SEVEN UFO (PROBABLY BALLOONS) MOVED FROM SW OF [redacted] TOWARD WEST SLOWLY AND PASSED [redacted] AND [redacted] ALT 54,ØØØ-69,ØØØ FT. C. Ø65Ø-Ø716, TWO UFO (PROBABLY BALLOONS) MOVED FROM NORTH OF [redacted] TOWARD WEST SLOWLY AND PASSED [redacted], ALT 1Ø,ØØØ FT. D.

TOP SECRET UMBRA

37

[redacted]

Ø653-Ø653, ONE ~~UFO~~-(PROBABLY A BALLOON) MOVED NORTH OF [redacted]

NNN

[redacted]

38

[redacted]

CATEGORY = 4ØØ   M = 67

MESSAGE = Ø3668915

LH #ØØ11 236Ø556
ZNY MMNSH
ZKZK PP [redacted] DE
[redacted]
ZEM
[redacted] SECTION EIGHT [redacted] THE
FOLLOWING INFORMATION HAS BEEN PROVIDED [redacted]

7. RADAR TRACKING OF UNIDENTIFIED FLYING OBJECTS (UFO: A. Ø345-Ø627, 14 UFO (PROBABLY BALLOONS) MOVED FROM SW OF [redacted] TOWARD WEST AND PASSED SE OF [redacted] AND EAST OF [redacted] ALT 1Ø,ØØØ FT. B. 1431-165Ø, TWO UFO (PROBABLY BALLOONS) MOVED OVER [redacted] AND [redacted] ALT 67,ØØØ-7Ø,ØØØ FT.

[redacted]

39

[redacted]

CATEGORY = 4ØØ M = 65

MESSAGE = Ø3668879

LH #ØØ11 2330756
ZNY MMNSH
ZKZK PP [redacted] DE
[redacted]
ZEM
[redacted] WNINTEL SECTION EIGHT [redacted] THE
FOLLOWING INFORMATION HAS BEEN PROVIDED [redacted]

[redacted]

40

T [redacted]

RADAR

TRACKING OF UNIDENTIFIED FLYING OBJECTS: 22.
#ØØ11 NNN

T [redacted]

41

[redacted]

CATEGORY = 4ØØ M = 66

MESSAGE = Ø366388Ø

LH #ØØ12 233Ø756
ZNY MMNSH
ZKZK PP [redacted] DE
[redacted]
ZEM
[redacted] SECTION NINE [redacted] THE
FOLLOWING INFORMATION HAS BEEN PROVIDED [redacted]

[redacted]

42

[redacted]

(UNIDENTIFIED FLYING OBJECTS: 34 [redacted] RE. RADAR TRACKING OF

#0012 NNN

[redacted]

43

[redacted]

CATEGORY = 4ØØ M = 64

MESSAGE = Ø365867Ø

LH #ØØØ9 23ØØ655
ZNY MMNSH
ZKZK PP [redacted] DE
[redacted]
ZEM
[redacted] WNINTEL SECTION SIX [redacted] THE
FOLLOWING INFORMATION HAS BEEN PROVIDED [redacted]

5. RADAR TRACKING OF UNIDENTIFIED FLYING OBJECTS (UFO): 1Ø48-154Ø, 21 UFO (PROBABLY BALLOONS) MOVED OVER [redacted] AND [redacted] ALT 69,ØØØ FT. [redacted] UNIDENTIFIED FLYING OBJECTS (UFO): [redacted]4Ø, 25 UFO (PROBABLY BALLOONS) MOVED FROM WEST OF [redacted] AND WEST OF [redacted] TOWARD WEST AND PASSED [redacted] AND [redacted] ALT 1Ø,ØØØ-69,ØØØ FT.

[redacted]

44

[redacted]

CATEGORY = 4ØØ M = 63

MESSAGE = Ø365866Ø

LH #ØØ14 229Ø756
ZNY MMNSH
ZKZK PP [redacted] DE
[redacted]
ZEM
S E C R E T SPOKE [redacted] WNINTEL SECTION NINE [redacted] THE
FOLLOWING INFORMATION HAS BEEN PROVIDED [redacted]

6. [redacted]
UNIDENTIFIED FLYING OBJECTS(UFO: Ø627-Ø8Ø5, 13 UFO (PROBABLY BALLOONS) MOVED SLOWLY FROM [redacted] TOWARD SW AND PASSED [redacted] AND [redacted] ALT 16,5ØØ FT.

[redacted]

45

[redacted]

CATEGORY = 4ØØ M F 62

MESSAGE = Ø3658644

LH #ØØ1Ø 228Ø732
ZNY MMNSH
ZKZK PP [redacted] DE
[redacted]
ZEM
[redacted] [redacted]TEL SECTION EIGHT [redacted] THE
FOLLOWING INFORMATION HAS BEEN PROVIDED [redacted]

T [redacted]

46

~~TOP SECRET UMBRA~~

7. RADAR TRACKING OF UNIDENTIFIED FLYING OBJECTS (UFO): 1349-1620, 20 UFO (PROBABLY BALLOONS) MOVED OVER [redacted] AND [redacted] ALT 69,000 FT. 8. [redacted] UNIDENTIFIED FLYING OBJECTS (UFO: E2, IMPDET.
#0010 NNN

[redacted]

47

[redacted]

CATEGORY = 4ØØ M = 61

MESSAGE = Ø3658632

LH #ØØ12 227Ø723
ZNY MMNSH
ZKZK PP [redacted] DE
[redacted]
ZEM
S[redacted] W[redacted]TEL SECTION SEVEN [redacted] THE
FOLLOWING INFORMATION HAS BEEN PROVIDED [redacted]

7. RADAR TRACKING OF UNIDENTIFIED FLYING OBJECTS (UFO): A. 1151-1331, 16 UFO (PROBABLY BALLOONS) MOVED SLOWLY FROM [redacted] AND [redacted] TOWARD WEST, ALT 49,2ØØ-75,5ØØ FT. B. 15Ø2-1515, ONE UFO (PROBABLY A BALLOON) MOVED FROM [redacted] TOWARD WEST, ALT 72,1ØØ FT. 8[redacted] UNIDENTIFIED FLYING OBJECTS (UFO): A. ØØ23-Ø71Ø, 16 UFO (PROBABLY BALLOONS) MOVED SLOWLY FROM NW OF [redacted] AND PASSED [redacted]

~~T O P S E C R E T U M B R A~~

48

T O P [redacted]

[redacted] AND [redacted] ALT 85,3ØØ-88,5ØØ FT. B. Ø528-1Ø38, 12 UFO (PROBABLY BALLOONS) MOVED SLOWLY FROM SW OF [redacted] TOWARD WEST AND PASSED [redacted] AND [redacted] ALT 62,4ØØ-72,1ØØ FT. C. Ø6Ø3-1Ø38, TWO UFO (PROBABLY BALLOONS) MOVED FROM SW OF [redacted] TOWARD WEST AND PASSED [redacted] ALT 1Ø,ØØØ FT. D. 114Ø-132Ø SIX UFO (PROBABLY BALLOONS) MOVED SLOWLY FROM [redacted] TOWARD NW AND PASSED [redacted] AND [redacted] ALT 1Ø,ØØØ-64,ØØØ FT. E. 1313-1332, THREE UFOS (PROBABLY BALLOONS) MOVED SLOWLY FROM [redacted] TOWARD WEST, ALT 3Ø,ØØØ- 36,ØØØ FT.

T O P [redacted]

49

[redacted]

CATEGORY = 4ØØ M = 6Ø

MESSAGE = Ø3658622

LH #ØØ1Ø 226Ø814
ZNY MMNSH
ZKZK PP [redacted] DE
[redacted]
ZEM
S E C R E T SPOKE [redacted] WNINTEL FINAL SECTION OF SEVEN [redacted]
THE FOLLOWING INFORMATION HAS BEEN PROVIDED [redacted]

RADAR TRACKING OF UNIDENTIFIED FLYING OBJECTS: 22. [redacted]

[redacted]

50

[redacted] U M B R A

CATEGORY = 4ØØ M = 59

MESSAGE = Ø3658610

LH #ØØ12 223Ø6Ø3
ZNY MMNSH
ZKZK PP [redacted] DE
[redacted]
ZEM
[redacted] SECTION NINE [redacted] THE
FOLLOWING INFORMATION HAS BEEN PROVIDED [redacted]

8. RADAR
TRACKING OF UNIDENTIFIED FLYING OBJECTS (UFO): A. Ø245-Ø85Ø, EIGHT UFO'S
(PROBABLY BALLOONS) MOVED SLOWLY FROM SW OF [redacted] TOWARD WEST AND PASSED

[redacted] U M B R A

51

[redacted] AND [redacted] ALT 49,2ØØ-62,4ØØ FT. B. Ø24Ø-Ø85Ø, ONE UFO (PROBABLY A BALLOON) MOVED SLOWLY FROM SOUTH OF [redacted], TURNED SW AFTER PASSING EAST OF [redacted] TURNED WEST UPON REACHING SOUTH OF [redacted] AND PASSED [redacted] AND [redacted], ALT 62,4ØØ FT. C. Ø713-141Ø, EIGHT UFO'S (PROBABLY BALLOONS) MOVED SLOWLY FROM [redacted] TOWARD WEST AND PASSED [redacted] ALT 75,5ØØ FT.

[redacted]

CATEGORY = 4ØØ M = 53

MESSAGE = Ø3653596

LH #ØØ12 222Ø636
ZNY MMNSH
ZKZK PP [redacted] DE
[redacted]
ZEM
[redacted] SECTION FIVE [redacted] THE
FOLLOWING INFORMATION HAS [redacted] PROVIDED [redacted]

[redacted]

53

~~TOP SECRET UMBRA~~

[redacted] 6. RADAR TRACKING OF UNIDENTIFIED FLYING OBJECTS (UFO): A. ØØ1Ø-Ø455, 11 UFO (PROBABLY BALLOONS) MOVED FROM [redacted] AND WEST OF [redacted] TOWARD WEST, PASSED SOUTH OF [redacted] AND ENTERED THE TERRITORIES OF [redacted] AND [redacted] B. Ø82Ø-1135, 11 UFO(PROBABLY BALLOONS) MOVED FROM [redacted] AND WEST OF [redacted] TOWARD WEST, PASSED SOUTH OF [redacted] AND ENTERED THE TERRITORIES OF [redacted]. C. Ø74Ø-Ø852, 15 UFO (PROBABLY BALLOONS) MOVED OVER [redacted] ALT 2Ø,ØØØ-29,8ØØ FT. D. 1345-162Ø, 11 UFO (PROBABLY BALLOONS) MOVED OVER [redacted] AND [redacted] ALT 59,ØØØ- 69,ØØØ FT. E2, IMPDET. #ØØ12 NNN

~~TOP SECRET UMBRA~~

54

~~TOP SECRET UMBRA~~

CATEGORY = 4ØØ M = 57

MESSAGE = Ø3658574

LH #ØØ14 22ØØ642
ZNY MMNSH
ZKZK PP [redacted] DE
[redacted]
ZEM
~~S E C R E T SPOKE~~ [redacted] INTEL SECTION SIX [redacted] THE
FOLLOWING INFORMATION HAS BEEN PROVIDED [redacted]

6. [redacted]
UNIDENTIFIED FLYING OBJECTS (UFO): ØØ11-Ø74Ø [redacted] RADAR TRACKED FOUR UFO
(PROBABLY BALLOONS) MOVING FROM BETWEEN [redacted] AND [redacted] TOWARD THE WEST,
PASSING [redacted] AND [redacted] AND ENTERING [redacted]
[redacted] ALT 72,1ØØ-75,5ØØ FT.

7. RADAR TRACKING OF UNIDENTIFIED
FLYING OBJECTS (UFO): (A. 1Ø4Ø-1446) TWO UFO (PROBABLY BALLOONS) MOVED OVER
[redacted] ALT 72,ØØØ-79,ØØØ FT, SPEED 41 KTS. (B. Ø846-1544)
EIGHT UFO (PROBABLY BALLOONS) MOVED OVER [redacted] AND [redacted]
[redacted], ALT 69,ØØØ-86,ØØØ FT.

~~TOP SECRET UMBRA~~

55

CATEGORY = 4ØØ   M = · 55

MESSAGE = Ø3658563

LH #ØØØ7 219Ø54Ø
ZNY MMNSH
ZKZK PP [redacted] DE
[redacted]
ZEM
S E C R E T [redacted] WNINTEL SECTION SEVEN [redacted] THE
FOLLOWING INFORMATION HAS BEEN PROVIDED [redacted]

RADAR TRACKING OF UNIDENTIFIED FLYING OBJECTS: (1).

56

~~TOP SECRET UMBRA~~

CATEGORY = 4ØØ M = 56

MESSAGE = Ø3658564

LH #ØØØ8 219Ø54Ø
ZNY MMNSH
ZKZX PP [redacted] DE
[redacted]
ZEM
~~S E C R E T~~ [redacted] WHI[redacted]EL FINAL SECTION OF EIGHT [redacted]
THE FOLLOWING INFORMATION HAS BEEN PROVIDE[redacted]

F. RADAR TRACKING OF UNIDENTIFIED FLYING OBJECTS: 14. G. [redacted]
UNIDENTIFIED FLYING OBJECTS: [redacted] 23 UFO'S.

~~TOP SECRET UMBRA~~

57

[redacted]

CATEGORY = 4ØØ M = 54

MESSAGE = Ø3653555

LH #ØØ12 216Ø549
ZNY MMNSH
ZKZK PP [redacted] DE
P Ø4Ø531Z AUG 78
[redacted]
ZEM
[redacted] UNINTEL SECTION EIGHT [redacted] THE
FOLLOWING INFORMATION HAS BEEN PROVIDED [redacted]

7. RADAR TRACKING OF UNIDENTIFIED FLYING OBJECTS (UFO): A. Ø1ØØ-Ø51, 17 UFO (PROBABLY BALLOONS) MOVED FROM WEST OF [redacted], ALT 82,ØØØ-88,2ØØ FT. B. Ø756-145Ø, 23 UFO (PROBABLY BALLOONS) MOVED OVER [redacted] ALT 66,ØØØ-71,5ØØ FT. C. Ø743-1523, FOUR UFO (PROBABLY BALLOONS) MOVED FROM [redacted]

[redacted]

58

[redacted]

CATEGORY = 4ØØ M = 53

MESSAGE = Ø3648346

LH #ØØØ5 215Ø544
ZNY MMNSH
ZKZK PP [redacted] DE
[redacted]
ZEM
S E C R E T SPOKE [redacted] UNINTEL SECTION FOUR CITE [redacted] THE
FOLLOWING INFORMATION HAS BEEN PROVIDED [redacted]

5. RADAR TRACKING OF UNIDENTIFIED FLYING OBJECTS (UFO). A. 1633-174Ø, FIVE UFO (PROBABLY BALLOONS) MOVED FROM [redacted] TOWARD WEST, PASSED [redacted] AND FADED [redacted], ALT 6,5ØØ-13,1ØØ FT. B. Ø848-16Ø9, 14 UFO (PROBABLY BALLOONS) MOVED FROM [redacted] TOWARD NW AND PASSED [redacted] ALT 72,1ØØ- 82,ØØØ FT.

T O P S E C R E T U M B R A

59

[redacted]

CATEGORY = 400 M = 52

MESSAGE = 03648331

LH #0010 2130540
ZNY MMNSH
ZKZK PP [redacted] DE
[redacted] #7

ZEM
[redacted] INTEL SECTION THREE [redacted] THE
FOLLOWING INFORMATION HAS BEEN PROVIDED [redacted]

5. [redacted] UNIDENTIFIED FLYING OBJECTS (UFO): 0655-0905, EIGHT UFO MOVED SLOWLY FROM [redacted] TOWARD SW AND PASSED [redacted] AND [redacted] ALT 11,500-72,100 FT.

6. RADAR TRACKING OF UNIDENTIFIED FLYING OBJECTS (UFO): A. 0050-0300, SEVEN UFO (PROBABLY BALLOONS) MOVED SLOWLY FROM NW OF [redacted] TOWARD WEST AND PASSED [redacted] ALT 85,300-88,500 FT. B. 0529-0710, 14 UFO (PROBABLY BALLOONS) MOVED SLOWLY FROM NE OF [redacted] TOWARD

[redacted]

60

[redacted]

WEST AND PASSED [redacted] ALT 8,2ØØ-13,1ØØ FT. C. 1Ø21-14Ø1, 14
UFO (PROBABLY BALLOONS) MOVED SLOWLY FROM [redacted] TOWARD WEST AND
PASSED [redacted] ALT
66,[illegible] FT. D. Ø719-Ø924, 12 UFO (PROBABLY BALLOONS) MOVED SLOWLY FROM [redacted] TOWARD SW AND FADED [redacted]
11Ø[illegible] ALT 57,ØØØ- 72,1ØØ FT. E. Ø749-141Ø, FOUR UFO (PROBABLY BALLOONS)
MOVED FROM [redacted] TOWARD
WEST. ALT 69,ØØØ-95,ØØØ FT. E. Ø719-Ø815, ONE UFO (PROBABLY A BALLOON) MOVED
FROM [redacted] TOWARDS WEST, ALT. 72,ØØØ-75,ØØØ FT.

[redacted] U M B R A

61

[redacted]

CATEGORY = 4ØØ M = 5Ø

MESSAGE = Ø3648325

LH #ØØ13 212Ø54Ø
ZHY MMNSH
ZKZK PP [redacted] DE
[redacted]
ZEM
[redacted] WNI[redacted] SECTION 5 CITE [redacted] THE FOLLOWING
INFORMATION HAS BEEN PROVI[redacted]

[redacted]

62

TOP SECRET UMBRA

5. RADAR TRACKING OF UNIDENTIFIED FLYING OBJECTS (UFO): A. ØØ35-Ø63Ø, 24 UFO (PROBABLY BALLOONS) MOVED FROM NW OF [redacted]OWARD WEST AND PASSED OVER [redacted], ALT 33,ØØØ-85,3ØØ FT. B. 124Ø-1321, 21 UFO (PROBABLY BALLOONS) MOVED FROM NW OF [redacted] TOWARD WEST AND PASSED OVER [redacted] ALT UNK. C. Ø725-1116, 14 UFO (PROBABLY BALLOONS) OVER [redacted] ALT 95,ØØØ FT. D. Ø815-1542, 28 UFO (PROBABLY BALLOONS) MOVED OVER [redacted] ALT 66,ØØØ-88,ØØØ FT. E. 1546-1625, ONE UFO (PROBABLY A BALLOON) MOVED OVER [redacted] ALT 75,5ØØ FT.

TOP SECRET UMBRA

63

[redacted]

CATEGORY = 4ØØ   M = 51

MESSAGE = Ø3648327

LH #ØØ15 212Ø54Ø
ZNY MMNSH
ZKZK PP [redacted] DE
[redacted]
ZEM
~~S E C R E T SPOKE~~ [redacted] ~~UNINTEL~~ SECTION 7 CITE [redacted] THE FOLLOWING
INFORMATION HAS BEEN PROVIDED [redacted]

F. RADAR TRACKING OF UNIDENTIFIED FLYING OBJECTS: ONE.

T O P [redacted] U M B R A

64

[redacted]

CATEGORY = 4ØØ M = 49

MESSAGE = Ø3648311

LH #ØØØ9 2ØØØ614
ZNY MMNSH
ZKZK PP [redacted] DE
[redacted]
ZEM
S E C R E T SPOKE [redacted] INTEL SEC 7 [redacted] THE FOLLOWING
INFORMATION HAS BEEN PROVIDED [redacted]

5. RADAR TRACKING OF UNIDENTIFIED FLYING OBJECTS (UFO): A. ØØ2Ø-1244, 31 UFO (PROBABLY BALLOONS) MOVED FROM BETWEEN NORTH OF [redacted] AND NE OF [redacted] TOWARD WEST SLOWLY, AND PASSED SOUTH OF [redacted] ALT 13,1ØØ-79,ØØØ FT. B. Ø93Ø-1443, TEN UFO (PROBABLY BALLOONS) MOVED FROM [redacted] TOWARD WEST SLOWLY, PASSED [redacted] AND FADED [redacted] AND [redacted] ALT 6Ø,ØØØ-72,1ØØ FT. C. 1ØØØ-1223, TWO UFO (PROBABLY BALLOONS) MOVED FROM [redacted] TOWARD WEST, ALT 79,ØØØ-1ØØ,ØØØ FT.

TOP SECRET UMBRA

65

T O P [redacted]

CATEGORY = 4ØØ M = 48

MESSAGE = Ø364830Ø

LH #ØØ11 2Ø8Ø625
ZNY MMNSH
ZKZK PP [redacted] DE
[redacted]
ZEM [redacted] SECTION SEVEN [redacted] THE
FOLLOWING INFORMATION HAS BEEN PROVIDED [redacted]

6. RADAR TRACKING OF UNIDENTIFIED FLYING OBJECTS (UFO): 1216-1234, FOUR UFO (PROBABLY BALLOONS MOVED FROM [redacted] TOWARD WEST, ALT 89,ØØØ FT. 7. [redacted] UNIDENTIFIED FLYING OBJECTS (UFO): A. Ø548-1ØØ4, FIVE UFO (PROBABLY BALLOONS) MOVED FROM WEST OF [redacted] TOWARD WEST, PASSED OVER [redacted] AND ENTERED [redacted], ALT 66,ØØØ FT. B. 1615-181Ø, 5Ø UFO (PROBABLY BALLOONS) MOVED FROM WEST OF [redacted] TOWARD WEST, PASSED OVER [redacted] AND ENTERED [redacted], ALT 6,5ØØ-2Ø,ØØØ [redacted] UFO
1731-1812, [redacted] ALT 15,ØØØ-23,ØØØ FT.

T O P [redacted]

66

[redacted]

CATEGORY = 400 M = 47

MESSAGE = 03648278

LH #0007 2060540
ZNY MMNSH
ZKZK PP [redacted] DE
[redacted]
ZEM
[redacted] SECTION 5 CITE [redacted] THE FOLLOWING INFORMATION HAS BEEN PROVIDED [redacted]

5. RADAR TRACKING OF UNIDENTIFIED FLYING OBJECT (UFO): 0754-0819, ONE UFO (PROBABLY A BALLOON) MOVED FROM [redacted] TOWARD WEST, ALT 63,000 FT. 6. [redacted]

T O P [redacted] U M B R A

67

UNIDENTIFIED FLYING OBJECTS (UFO): A. Ø5Ø8-Ø851, SIX UFO (PROBABLY BALLOONS) MOVED FROM NORTH OF [redacted] TOWARD WEST AND PASSED OVER [redacted], ALT 66,ØØØ-72,1ØØ FT.

68

[redacted]

CATEGORY = 42Ø M = 45

MESSAGE = Ø3648269

LH #ØØ11 2Ø5Ø6Ø1
ZNY MMNSH
ZKZK PP [redacted] DE
[redacted]
ZEM
[redacted] SECTION 7:CITE COS TAIPEI: 2112 THE FOLLOWING INFORMATION HAS BEEN PROVIDED [redacted]

F. RADAR TRACKING OF UNIDENTIFIED FLYING OBJECTS: I2.

[redacted]

69

CATEGORY = 400 M = 46

MESSAGE = 03648270

LH #0012 2050601
ZNY MMNSH
ZKZK PP [redacted] DE
[redacted]
ZEM
[redacted] EL SECTION 8 [redacted] THE FOLLOWING
INFORMATION HAS BEEN PROVIDED [redacted]

F. RADAR

TRACKING OF UNIDENTIFIED FLYING OBJECTS: 39.

[redacted]

70

TOP SECRET UMBRA

CATEGORY = 400 M` = [illegible] 43

MESSAGE = 03648255

LH #0016 2020713
ZNY MMNSH
ZKZK PP [redacted] DE
[redacted]
ZEM
S E C R E T SPOKE [redacted] .WNINTEL SEC 8 [redacted] THE FOLLOWING
INFORMATION HAS BEEN PROVIDED [redacted]

TOP SECRET UMBRA

71

TOP SECRET UMBRA

6. [redacted] UNIDENTIFIED FLYING OBJECTS (UFO: ØØ19-12ØØ, [redacted] RADAR TRACKED 27 UFO (PROBABLY BALLOONS) MOVING FROM [redacted] TOWARD WEST, PASSING SE OF [redacted] #ØØ16 NNN

TOP SECRET UMBRA

72

[redacted]

CATEGORY = 400 M = 44

MESSAGE = 03648256

LH #0017 2020713
ZNY MMNSH
ZKZK PP [redacted] .DE
[redacted]
ZEM
[redacted] WNI[redacted] SEC 9 [redacted] THE FOLLOWING
INFORMATION HAS BEEN PROVIDED [redacted]

7. RADAR TRACKING OF UNIDENTIFIED FLYING OBJECTS: A. 1018-1609, 18 UFO (PROBABLY BALLOONS) MOVED OVER [redacted] ALT 75,500-85,300 FT. B. 1408-1545, 22 UFO (PROBABLY BALLOONS) MOVED OVER [redacted] ALT 86,000 FT.

[redacted]

73

~~TOP SECRET UMBRA~~

0019-2000Z [redacted] RADAR TRACKED 33 FLTS OF UNKNOWN OBJECTS (PROBABLY BALLOONS) MOVING SLOWLY FROM THE VICINITY OF [redacted] ALT. 10,000-28,000 METERS.

[redacted] ET UMBRA

74

[redacted]

CATEGORY = 4ØØ M = 42

MESSAGE = Ø3648241

LH #2212 2Ø1Ø618
ZNY MMNSH
ZKZK PP [redacted] DE
[redacted]
ZEM
[redacted] E [redacted] WN[redacted]TEL SEC 8 [redacted] THE FOLLOWING INFORMATION HAS BEEN PROVIDED [redacted]

5. RADAR TRACKING OF UNIDENTIFIED FLYING OBJECTS (UFO): A. ØØ3Ø-Ø625, 25 UFO (PROBABLY BALLOONS) MOVED SLOWLY FROM OVER THE [redacted] BETWEEN SW OF [redacted] AND WEST OF [redacted]

[redacted]

75

~~TOP SECRET UMBRA~~

TOWARD WEST AND PASSED [redacted] ALT 2Ø,ØØØ-95,3ØØ FT. B.
1435-1651, TWO UFO (PROBABLY BALLOONS) MOVED FROM NW OF [redacted]
TOWARD WEST AND FADED EAST OF [redacted] ALT 62,4ØØ FT. C.
Ø745-Ø85Ø, 17 UFO (PROBABLY BALLOONS) MOVED FROM [redacted] TOWARD
WEST, ALT 95,3ØØ FT.

~~TOP SECRET UMBRA~~

76

[redacted]

CATEGORY = 4ØØ M = 41

MESSAGE = Ø3637534

LH #ØØ13 199Ø536
ZNY MMNSH
ZKZK PP [redacted] DE
[redacted]
ZEM
S E C R E T [redacted] WNINTEL SECTION SIX [redacted] THE
FOLLOWING INFORMATION HAS BEEN PROVIDED [redacted]

7. RADAR TRACKING OF UNIDENTIFIED FLYING OBJECTS (UFO): A. Ø45Ø-1Ø2Ø, 18 UFO (PROBABLY BALLOONS) MOVED SLOWLY FROM OVER [redacted] BETWEEN SW OF [redacted]

[redacted]

77

AND WEST OF [redacted] TOWARD WEST, PASSED [redacted], ALT 66,ØØØ-95,3ØØ FT. B. Ø848-1545, 16 UFO (PROBABLY BALLOONS) MOVED SLOWLY FROM [redacted] RESPECTIVELY TOWARD WEST AND FADED [redacted] ALT 69,ØØØ-79,ØØØ FT.

T [redacted]

CATEGORY = 4ØØ M = 38

MESSAGE = Ø3637519

LH #ØØØ8 198Ø557
ZNY MMNSH
ZKZK PP [redacted] DE
[redacted]
ZEM
[redacted] SECTION 4 [redacted] THE FOLLOWING
INFORMATION HAS BEEN PROVIDED [redacted]

[redacted] U M B R A

79

TOP SECRET UMBRA

5. [redacted]
UNIDENTIFIED FLYING OBJECTS (UFO: Ø42Ø-144Ø, 23 UFO (PROBABLY BALLOONS) MOVED FROM WEST OF [redacted] PASSED SW OF [redacted] AND ENTERED THE [redacted] ALT 66 ØØØ-69 ØØØ FT

6. RADAR TRACKING OF UNIDENTIFIED FLYING OBJECTS (UFO): A. Ø715-141Ø, EIGHT UFO (PROBABLY BALLOONS) MOVED OVER [redacted] ALT 66,ØØØ-79,ØØØ FT. B. Ø925-15Ø9, NINE UFO (PROBABLY BALLOONS) MOVED SLOWLY FROM [redacted] TOWARD WEST, PASSED [redacted] AND FADED NEAR [redacted] ALT 66,ØØØ-85,ØØØ FT. C. 1Ø46-1344, THREE UFO (PROBABLY BALLOONS) MOVED SLOWLY FROM [redacted] TOWARD WEST AND FADED NEAR [redacted] ALT 72,1ØØ-85,3ØØ FT. #ØØØ8 NNN

TOP SECRET UMBRA

80

[redacted]

CATEGORY = 430 M = 39

MESSAGE = 03637521

LH #0010 1980557
ZNY MMNSH
ZKZK PP [redacted] DE

[redacted]

ZEM

[redacted] WN[redacted] SECTION 6 [redacted] THE FOLLOWING
INFORMATION HAS BEEN PROVIDED [redacted]

F. RADAR TRACKING OF UNIDENTIFIED FLYING OBJECTS: 52.

T [redacted]

81

[redacted]

CATEGORY = 4ØØ M = 4Ø

MESSAGE = Ø3637522

LH #ØØ11 198Ø557
ZNY MMNSH
ZKZK PP [redacted] DE
[redacted]
ZEM
[redacted] NTEL FINAL SECTION [redacted] THE
FOLLOWING INFORMATION HAS BEEN PROVIDED [redacted]

F. RADAR TRACKING OF
UNIDENTIFIED FLYING OBJECTS: NINE.

T [redacted]

82

[redacted]

CATEGORY = 400 M = 37

MESSAGE = 03637510

LH 00010 1950643
ZNY MMNSH
ZKZK PP [redacted] DE
[redacted]
ZEM
[redacted] SECTION 8 [redacted] THE FOLLOWING
INFORMATION HAS BEEN PROVIDED [redacted]

T[redacted]

83

6. RADAR TRACKING OF <u>UNIDENTIFIED FLYING OBJECTS</u> (UFO):
1535-1640 FIVE UFO (PROBABLY BALLOONS) MOVED SLOWLY FROM SE OF [redacted] TOWARD WEST.

~~TOP SECRET UMBRA~~

84

[redacted]

CATEGORY = 4ØØ M = 36

MESSAGE = Ø3637499

LH #ØØ11 194Ø63Ø
ZNY MMNSH
ZKZK PP [redacted] DE
[redacted]

ZEM
[redacted] SECTION EIGHT [redacted] THE
FOLLOWING INFORMATION WAS BEEN PROVIDED [redacted]

6. [redacted]
REACTION TO UNIDENTIFIED FLYING OBJECTS (UFO): ØØ12-Ø73Ø, 14 UFO (PROBABLY BALLOONS) MOVED FROM [redacted] TOWARD WEST AND PASSED [redacted] ALT 6,5ØØ-85,3ØØ FT

7. RADAR TRACKING OF UNIDENTIFIED FLYING OBJECTS(UFO): 1645-1737, EIGHT UFO (PROBABLY BALLOONS) MOVED FROM [redacted] AND PASSED [redacted] ALT 18,ØØØ-29,2ØØ FT.

[redacted]

85

[redacted]

[redacted]

CATEGORY = 4ØØ M = 35

MESSAGE = Ø3637477

LH #ØØØ9 192Ø513
ZNY MMNSH
ZKZK PP [redacted].DE
[redacted]
ZEM
[redacted] SECTION FIVE [redacted] THE
FOLLOWING INFORMATION HAS BEEN PROVIDED [redacted]

T O P S E C R E T U M B R A

86

[redacted]

[redacted]
5. RADAR TRACKING OF
UNIDENTIFIED FLYING OBJECTS (UFO): A. Ø51Ø-Ø913, FIVE UFO (PROBABLY BALLOONS)
MOVED FROM NW OF [redacted], PASSED SE OF [redacted] ALT 56,ØØØ-66,ØØØ FT. E2, IMPDET. #ØØØ9 NNN

[redacted]

87

[redacted]

[redacted]

CATEGORY = 4ØØ　　　M = 34

MESSAGE = Ø363747Ø

LH #ØØ19 191Ø848
ZNY MMNSH
ZKZK PP [redacted] DE
[redacted]
ZEM
[redacted] FINAL SECTION OF SEVEN [redacted]
THE FOLLOWING INFORMATION HAS BEEN PROVIDED [redacted]

RADAR TRACKING OF UNIDENTIFIED FLYING OBJECTS: 14.

[redacted]

88

T O P S E C R E T U M B R A

CATEGORY = 4ØØ M = 33

MESSAGE = Ø3637454

LH #ØØ16 188Ø542
ZNY MMNSH
ZKZK PP [redacted] DE
[redacted]
ZEM
S E C R E T SPOKE [redacted] SECTION 8 [redacted] THE FOLLOWING
INFORMATION HAS BEEN PROVIDED [redacted]

-. RADAR TRACKING OF UNIDENTIFIED FLYING OBJECT (UFO:
1350-1404, ONE UFO (PROBABLY A BALLOON) MOVED NEAR [redacted], ALT
62,4ØØ FT. 8. [redacted] UNIDENTIFIED FLYING OBJECTS (UFO: A. Ø445-11Ø7.

T O P S E C R E T U M B R A

89

TOP SECRET UMBRA

THREE UFO (PROBABLY BALLOONS) MOVED OVER [redacted] ALT 25,ØØØ-52,5ØØ FT.

TOP SECRET UMBRA

90

~~T O P S E C R E T U M B R A~~

CATEGORY = 4ØØ M = 31

MESSAGE = Ø3637438

LH #ØØ14 187Ø516
ZNY MMNSH
ZKZK PP [redacted] DE
[redacted]
ZEM
~~S E C R E T SPOKE~~ [redacted] SECTION 9 [redacted] THE FOLLOWING
INFORMATION HAS BEEN PROVIDED [redacted]

5. RADAR
TRACKING OF UNIDENTIFIED FLYING OBJECTS(UFO): A. 1ØØØ-1Ø58, SEVEN UFO
(PROBABLY BALLOONS) MOVED SLOWLY FROM SOUTH OF [redacted] TOWARD WEST AND PASSED
[redacted] ALT 85,3ØØ FT. B. 1212-13Ø5, TWO UFO
(PROBABLY BALLOONS) MOVED SLOWLY FROM NORTH OF [redacted] WEST AND FADED NEAR
[redacted]

~~T O P S E C R E T U M B R A~~

91

T[redacted]

CATEGORY = 4ØØ M = 32

MESSAGE = Ø363744Ø

LH #ØØ16 187Ø516
ZNY MMNSH
ZKZK PP [redacted].DE
[redacted]
ZEM
S[redacted] FINAL SECTION OF 11[redacted] THE
FOLLOWING INFORMATION HAS BEEN PROVIDED[redacted]

UNIDENTIFIED FLYING OBJECTS: THREE

F. RADAR TRACKING OF

T[redacted]

92

[redacted]

T[redacted]S E C R E T U M B R A

CATEGORY = 4ØØ M = 97

MESSAGE = Ø3749165

LH #ØØ1Ø 36ØØ619
ZNY MMNSH
ZKZK PP [redacted] DE
[redacted]
ZEM
[redacted] SECTION 7 [redacted] THIS MESSAGE IS CLASSIFIED SECRET SPOKE IN ITS ENTIRETY. THE FOLLOWING INFORMATION HAS BEEN PROVIDED [redacted]

[redacted]

93

TOP SECRET UMBRA

I. RADAR TRACKING OF [UNIDENTIFIED FLYING OBJECTS: ONE].

TOP SECRET UMBRA

94

TOP SECRET UMBRA

DE [redacted] #3971 0752123
ZNY MMNSH
ZKZK RR [redacted] DE
[redacted]
FM DIRNSA
[redacted]
ZEM
T O P S E C R E T UMBRA
XXMMENP01F0079074
3/00/9953-79
[redacted]
XXCC
(THIS PRODUCT IS CLASSIFIED TOP SECRET UMBRA IN ITS ENTIRETY)
[redacted] TOOK PLACE SIX
MONTHS AGO IN THE [redacted] REGION, AND THE
INCREASE IN FLYING SAUCER ACTIVITY, WHICH HE EXPLAINED AS HAVING
A CONNECTION WITH THE PLAN.
[redacted]
XXHH
MS99737
292
#3971
NNN

TOP SECRET UMBRA

95

[redacted]

CATEGORY = 400 M = 11

MESSAGE = 03799668

LH #0008 0730717
ZNY MMNSH
ZKZK PP [redacted] DE

[redacted] WHINTEL SECTION SIX, [redacted] THIS
MESSAGE IS CLASSIFIED SECRET SPOKE IN ITS ENTIRETY. THE FOLLOWING INFORMATION
HAS BEEN PROVIDED [redacted]

[redacted]

96

[redacted]

RADAR TRACKING OF UNIDENTIFIED FLYING OBJECT (UFO): [redacted] #ØØØ8 NNN

[redacted]

97

[redacted]

CATEGORY = 4ØØ M = 6

MESSAGE = Ø3769685

LH #ØØ16 Ø4ØØ917
ZNY MMNSH
ZKZK PP [redacted] DE [redacted]

ZEM
[redacted] WNINTEL SECTION 8 [redacted] THIS MESSAGE IS CLASSIFIED SECRET SPOKE IN ITS ENTIRETY. THE FOLLOWING INFORMATION HAS BEEN PROVIDED [redacted]

7. RADAR TRACKING OF UNIDENTIFIED FLYING OBJECT (UFO: 1424-1539, ONE UFO (PROBABLY A BALLOON) MOVED SLOWLY FROM 43 NAUTICAL MILES (NM) SE OF [redacted] TOWARD EAST, PASSED [redacted] AND [redacted] AND FADED 16 NM SOUTH OF [redacted] ALT 43,6ØØ-49,2ØØ FT. 8. [redacted] REACTION TO UFO: 12Ø9-1455, ONE UFO (PROBABLY A BALLOON) MOVED SLOWLY FROM [redacted] TOWARD EAST, PASSED [redacted] AND [redacted], AND FADED [redacted] ALT 33,ØØØ-46,ØØØ FT [redacted] REACTION: 1241-1353, SIX [redacted] DEPARTED [redacted] FOR ATTACKING SAID UFO OVER [redacted] AND [redacted] ALT 26,2ØØ FT, SPEED 43Ø KTS.

[redacted]

98

TOP SECRET UMBRA

CATEGORY = 400 M = 5

MESSAGE = 03769658

LH #0012 0380630
ZNY MMNSH
ZKZK PP [redacted] DE
[redacted]
ZEM
S E C R E T SPOKE [redacted] [redacted] SECTION 10 [redacted] THIS
MESSAGE IS CLASSIFIED SECRET SPOKE IN ITS ENTIRETY. THE FOLLOWING INFORMATION
HAS BEEN PROVIDED: [redacted]

[redacted]

99

6. RADAR TRACKING OF [UNIDENTIFIED FLYING OBJECT (UFO): Ø822-11Ø3, ONE UFO (PROBABLY A BALLOON) MOVED SLOWLY FROM 124 NAUTICAL MILES (NM) 77 DEGREES (DEG) OFF [redacted] 11Ø16) [TOWARD EAST AND FADED 157 NM 1Ø5 DEG OFF [redacted] ALT 39,4ØØ-59,ØØØ FT.]

100

[redacted]

CATEGORY = 4ØØ M = 4

MESSAGE = Ø3769621

LH #ØØØ8 Ø33Ø653
ZNY MMNSH
ZKZK PP [redacted] DE
[redacted]
ZEM
[redacted] SECTION THREE [redacted]
THIS MESSAGE IS CLASSIFIED SECRET SPOKE IN ITS ENTIRETY. THE FOLLOWING
INFORMATION HAS BEEN PROVIDED [redacted]

[redacted]

101

~~TOP SECRET UMBRA~~

ste 5. RADAR TRACKING OF UNIDENTIFIED FLYING OBJECTS (UFO): Ø8Ø4-Ø834, SEVEN UFO (PROBABLY BALLOONS) MOVED OVER [redacted] ALT 11,5ØØ-16,ØØØ FT. XXHH #ØØØ8 NNN

~~TOP SECRET UMBRA~~

162

~~TOP SECRET UMBRA~~

CATEGORY = 4ØØ　　　M = 8

MESSAGE = Ø3777584

LN #ØØ22 Ø22Ø922
ZNY MMORN
ZKZK RR [redacted] DE
[redacted]
ZEM.
[redacted] WARNING NOTICEINTELLIGENCE SOURCES AND METHODS INVOLVED [redacted] THE FOLLOWING INFORMATION HAS BEEN RECEIVED [redacted] A [redacted] TWO [redacted] AGAINST AN UNIDENTIFIED FLYING OBJECT, BELIEVED TO HAVE BEEN A BALLOON, IN FLIGHT SOUTH OF [redacted]

~~TOP SECRET UMBRA~~

163

~~T O P S E C R E T U M B R A~~

CATEGORY = 400 - M = 3

MESSAGE = 03758369

LH #0016 0110644
ZNY MMNSH
ZKZK PP [redacted] DE
[redacted]
ZEM
~~S E C R E T SPOKE~~ [redacted] ~~UNINTEL~~ SECTION SEVEN, [redacted]. THIS MESSAGE IS CLASSIFIED SECRET SPOKE IN ITS ENTIRETY. THE FOLLOWING INFORMATION HAS BEEN PROVIDED [redacted]

[redacted]

104

[redacted]

[redacted]5. RADAR TRACKING OF UNIDENTIFIED FLYING OBJECT (UFO): Ø927-Ø935, ONE UFO (PROBABLY A BALLOON) MOVED OVER [redacted] IN [redacted] ALT 49,2ØØ FT.

T[redacted]

105

[redacted]

CATEGORY = 400 M = 2

MESSAGE = 03758358

LH #0015 0100713
ZNY MMNSH
ZKZK PP [redacted] DE
[redacted]
ZEM
S E C R E T SPOKE [redacted] UNINTEL SECTION TEN [redacted] THIS MESSAGE IS CLASSIFIED SECRET SPOKE IN ITS ENTIRETY. THE FOLLOWING INFORMATION HAS BEEN PROVIDED [redacted]

5. RADAR TRACKING OF UNIDENTIFIED FLYING OBJECT (UFO): 0914-0936, ONE UFO (PROBABLY A BALLOON) MOVED SLOWLY FROM THE WATERS THREE NM SOUTH OF [redacted] TOWARD EAST, ALT 36,000-49,200 FT.

[redacted]

106

14 H·15

[redacted]

CATEGORY = 4ØØ M = 1

MESSAGE = Ø3758292

LH #ØØØ3 ØØ3Ø614
ZNY MMNSH
ZKZK PP [redacted] DE
[redacted]
ZEM [redacted] UNINTEL SECTION 4 [redacted] THIS MESSAGE IS CLASSIFIED SECRET SPOKE IN ITS ENTIRETY. THE FOLLOWING INFORMATION HAS BEEN PROVIDED [redacted]

RADAR TRACKING OF UNIDENTIFIED FLYING OBJECTS: TWO

[redacted]

107

[redacted]

CATEGORY = 400 M = 93

MESSAGE = 03748935

LN #0633/1 3630928
ZNY MMORN
ZKZK RR [redacted] DE
[redacted]
ZEM
[redacted] NOFORN SECTION 1 OF 2 [redacted] WARNING NOTICE-INTELLIGENCE SOURCES AND METHODS INVOVLED [redacted] THE FOLLOWING INFORMATION HAS BEEN RECEIVED [redacted]

[redacted]

108

C. ON [redacted] THREE
[redacted] TWO DURING 1404 THROUGH 1440 AND ONE
DURING 1349 THROUGH 1549, [redacted] AN UNIDENTIFIED FLYING OBJECT
(PROBABLY A BALLOON) FLYING OVER [redacted] THE [redacted]
[redacted] DURING 1349 THROUGH 1549 [redacted] DUE TO BAD WEATHER AT
[redacted]. #0633 NNN

[redacted]

109

TOP SECRET [redacted]

CATEGORY = 4ØØ M = 92

MESSAGE = Ø374Ø57Ø

LH #ØØ14 348Ø647
ZNY MMNSH
ZKZK PP [redacted] DE
[redacted]
ZEM
[redacted] SECTION EIGHT, [redacted] THIS MESSAGE IS CLASSIFIED SECRET SPOKE IN ITS ENTIRETY. THE FOLLOWING INFORMATION HAS BEEN PROVIDED [redacted] THIS MESSAGE IS CLASSIFIED SECRET SPOKE IN ITS ENTIRETY.

56. RADAR TRACKING OF UNIDENTIFIED FLYING OBJECTS: (UFO): 1Ø38-1Ø56, FOUR UFO (PROBABLY BALLOONS) MOVED OVR [redacted] AND [redacted], ALT 79,ØØØ- 83,ØØØ FT.

[redacted] SECRET [redacted]

110

[redacted]

CATEGORY = 4ØØ M = 91

MESSAGE = Ø374Ø519

LH #ØØØ9 345Ø724
ZNY MMNSH
ZKZK PP [redacted] DE
[redacted]
ZEM
[redacted] SECTION 7 [redacted] THIS MESSAGE
IS CLASSIFIED SECRET SPOKE IN ITS ENTIRETY. THE FOLLOWING INFORMATION HAS BEEN
PROVIDED [redacted]

[redacted]

111

6. RADAR TRACKING OF UNIDENTIFIED FLYING OBJECTS (UFO:) A. 1Ø43-1133, ONE UFO (PROBABLY A BALLOON) FLEW FROM 1Ø5 NAUTICAL MILES (NM) 284 DEGREES (DEG) OFF [redacted] AND FADED 128 NM 2929 DEG OFF [redacted], ALT 5ØØ-59,ØØØ FT. B. 1229-1236, ONE UFO (PROBABLY A BALLOON) FLEW FROM 1Ø5 NM 284 DEG OFF [redacted] AND FADED 128 NM 2929 DEG OFF [redacted]. ALT 49,2ØØ-56,ØØØ FT.

112

T[redacted]

CATEGORY = 4ØØ M = 9Ø

MESSAGE = Ø373Ø421

LH #ØØ15 328Ø7Ø6
ZNY MMNSH
ZKZK PP[redacted] DE
[redacted]
ZEM
S E C R E [redacted] SECTION TEN [redacted] THE
FOLLOWING INFORMATION HAS BEEN PROVIDED [redacted]

5. RADAR TRACKING OF UNIDENTIFIED FLYING OBJECT (UFO):
22 NOV 1539-1546, ONE UFO (PROBABLY A BALLOON) MOVED 42 NAUTICAL MILES WEST OF [redacted] ALT 28,ØØØ FT. (DELAYED INFORMATION)

T[redacted]

113

[redacted]

CATEGORY = 8 M = 1

MESSAGE = Ø37ØØ614

LH #ØØØ2 276Ø7Ø7
ZNY MMNSH
ZKZK PP [redacted] DE
[redacted]
ZEM
S E C R E T [redacted] SECTION ONE OF TWO [redacted] THE
FOLLOWING INFORMATION HAS BEEN PROVIDED [redacted]

T [redacted]

114

[redacted]

5. RADAR TRACKING OF UNIDENTIFIED FLYING OBJECTS (UFO): A. Ø542-1114, FIVE UFO (PROBABLY BALLOONS) MOVED #ØØØ2 NNN

[redacted]

115

CATEGORY = 8 M = 2

MESSAGE = Ø37ØØ615

LH #ØØØ3 276Ø7Ø7
ZNY MMNSH
ZKZK PP [redacted] DE
[redacted]

ZEM
S E C R E T [redacted] SECTION TWO FINAL [redacted] THE
FOLLOWING INFORMATION HAS BEEN PROVIDED [redacted]

1ØØ9-154Ø, THREE UFO (PROBABLY BALLOONS) MOVED OVER [redacted] ALT
27,ØØØ-75,5ØØ FT.

116

[redacted]

CATEGORY = 400 M = 71

MESSAGE = 03668979

LH #0021 2430801
ZNY MMNSH
ZKZK PP [redacted] DE
[redacted]

ZEM
S E C R E T SPOKE [redacted]EL SECTION TEN [redacted] THE
FOLLOWING INFORMATION HAS BEEN PROVIDED [redacted]

6. RADAR TRACKING OF UNIDENTIFIED FLYING OBJECTS (UFO): A. 29 AUG 2300-0330, TEN UFO (PROBABLY BALLOONS) MOVED SLOWLY FROM WEST OF [redacted] TOWARD WEST AND PASSED [redacted] AND [redacted], ALT 66,000-85,300 FT. B. 0256-0505, 12 UFO (PROBABLY BALLOONS) MOVED SLOWLY FROM [redacted] TOWARD SOUTH AND PASSED [redacted] AND [redacted], ALT 10,000-15,000 FT. C. 0600-0839, SEVEN UFO (PROBABLY BALLOONS) MOVED SLOWLY FROM WEST OF [redacted] TOWARD WEST AND PASSED SOUTH OF [redacted] AND [redacted] D. 0653-0830, 13 UFO (PROBABLY BALLOONS) MOVED OVER [redacted] ALT 88,000 FT.

T[redacted]

17

~~TOP SECRET UMBRA~~

CATEGORY = 4ØØ M = 7Ø

MESSAGE = Ø3668953

LH #ØØ11 241Ø551
ZNY MMNSH
ZKZK PP [redacted] DE
[redacted]
ZEM
S E C R E T SPOKE [redacted] WHINTEL SECTION SEVEN [redacted] THE
FOLLOWING INFORMATION HAS BEEN PROVIDED [redacted]

5. RADAR TRACKING OF UNIDENTIFIED FLYING OBJECTS (UFO): A. Ø5Ø2-Ø94Ø, 46 UFO (PROBABLY BALLOONS) MOVED SLOWLY FROM NW OF [redacted] TOWARD WEST AND PASSED [redacted], ALT 13,1ØØ- 88,5ØØ FT. B. Ø6Ø5-Ø811, ONE UFO (PROBABLY A BALLOON) MOVED SLOWLY FROM [redacted] TOWARD WEST AND PASSED [redacted] AND [redacted] ALT 15,ØØØ-79,ØØØ FT. C. Ø7Ø9-Ø811, ONE UFO (PROBABLY A BALLOON) MOVED SLOWLY FROM [redacted] TOWARD WEST AND PASSED [redacted] ALT 15,ØØØ-79,ØØØ FT. D. 1634-1637, ONE UFO (PROBABLY A BALLOON) MOVED OVER [redacted] ALT 8Ø,ØØØ FT.

~~TOP SECRET UMBRA~~

118

T C P [redacted]

CATEGCRY = 4Ø5 M = 6

MESSAGE = Ø1229912

ZCZCØCLUØØØ3ØØØØØØØØØØØØØØØCFØØØ476Ø Ø
PP [redacted] ØCE
YACRFC Ø94 193Ø8ØØØ Ø
ZKZK PP [redacted] CE
[redacted]
ZEM
ØT O P S E C R E T [redacted] NCFCRN SECTICN CNE CF TWC [redacted] THE
FCLLCWING INFORMATICN HAS BEEN PRCVIDED [redacted]

TRACKING CF UNIDENTIFIEC FLYING CEJECTS ØØRADAR

[redacted]

119

[redacted]

CATEGORY = 405 . M = 9

MESSAGE = 01255651

ZCZCCRE099CRUM836
PP [redacted] CE
[redacted]
ZNY MMCRN
ZKZK [redacted] CE
[redacted]
ZEM
~~TOP SECRET UMBRA~~ [redacted] ~~NOFORN~~ FINAL SECTION OF TWO CITE 7637 3.

15. [redacted] RADAR TRACKING OF <u>UNIDENTIFIED FLYING OBJECTS</u>, A. 1649-1731, [redacted] RADAR TRACKED ONE UNIDENTIFIED FLYING OBJECT (UFO) MOVING OVER [redacted] AND [redacted] B. 0833-1101, [redacted] RADAR TRACKED ONE UFO MOVING OVER [redacted] C. 1503-1510, [redacted] RADAR TRACKED ONE UFO MOVING OVER 38-49 NM WEST OF [redacted] 3#0 #0142 NNNN [redacted]

~~T O P S E C R E T U M B R A~~

120

[redacted]

CATEGORY = 405 M = 11

MESSAGE = 01329574

ZCZCZKZK
RR [redacted] CE
PP [redacted] CE
ZNY MMNSH P [redacted]
FM DIRNSA
[redacted]
ZEM
[redacted] SECTION SEVEN
XXMM
[redacted]/C/[redacted]/R273-79 REPRODUCTION/FURTHER DISSEMINATION OF THIS DOCUMENT AS A WHOLE PROHIBITED WITHOUT PRIOR DIRNSA CONCURRENCE. SELECTED ITEMS MAY BE EXTRACTED AND FURTHER DISSEMINATED. [redacted] (IN PARAGRAPHS SO MARKED)

(C [redacted] AN UNIDENTIFIED FLYING OBJECT ALONG THE [redacted] NORTHEAST OF [redacted] A REFLECTED SPEED OF 54 KNOTS AND ALTITUDES RANGING FROM 11,480 TO 14,760 FEET INDICATE THE OBJECT MAY HAVE BEEN A BALLON.

[redacted]

121

[redacted]

CATEGORY = 405 M = 10

MESSAGE = Ø171Ø624

18Ø214111B1"1111764 #2135 NNNNVZ11CZCCLLS311CRC186
PP [redacted] CE
[redacted]
ZNY MMCRN
ZKZK PP [redacted] CE
[redacted]
ZEM
[redacted] NCFCRN [redacted]

[redacted]

122

[redacted]

7. [redacted] RADAR TRACKING OF UNIDENTIFIED FLYING OBJECT, 0736, [redacted] RADAR TRACKED ONE UNIDENTIFIED OBJECT MOVING OVER [redacted] FROM [redacted].

[redacted]

123

[redacted]

CATEGORY = 405 M = 6

MESSAGE = 01871297

ZCZCC03 491DRB982 LOG LN NO. 069
PP [redacted] CE
[redacted]
ZNY MMCRN
ZKZK PP [redacted] CE
[redacted]
ZEM
~~T O P S E C R E T UMBRA~~ [redacted] NOFORN CITE, 2049 FINAL SECTION OF FOUR

[redacted]

124

7. [redacted] RADAR TRACKING OF <u>UNIDENTIFIED FLYING CBJECTS</u> (UFO'S), A. 1Ø49-1Ø59, [redacted] RADAR TRACKED ONE UFO MOVING OVER 3Ø NM SCUTHEAST OF [redacted]. B. 11Ø6, [redacted] RADAR TRACKEC CNE UFO MOVING OVER 32 MN EAST CF [redacted] 89Ø #Ø3Ø5 NNNN [redacted]

125

[redacted]

CATEGORY = 405 M = 66

MESSAGE = 02659746

XO #0911 2840932
ZNY MMCRN
ZKZK RR [redacted] DE
[redacted]
ZEM
[redacted] WARNING NOTICE: SENSITIVE
INTELLIGENCE SOURCES AND METHODS INVOLVED [redacted] . THE FOLLOWING INFORMATION
HAS BEEN RECEIVED [redacted]
RADAR TRACKING OF UNIDENTIFIED OBJECT FLIGHT IS
[redacted] TRACKING OF UNIDENTIFIED OBJECT FLIGHT IN [redacted]
PPROXIMATELY 35 KILOMETERS NORTHWEST OF [redacted]
PLOTTED UNIDENTIFIED FLYING OBJECT POSITIONS ARE
APPROXIMATE). [redacted] THE FLIGHT OF AN UNIDENTIFIED OBJECT [redacted]
APROXIMATELY 35 KILOMETERS NORTHWEST OF [redacted]
THE UNIDENTIFIED OBJECT FLYING AT 120 KMH WAS INITIALLY DETECTED AT [redacted]
AT 1504H AND TRACKED EASTWARD TO [redacted] NORTH WARD TO
[redacted] E AT 1533 HRS AT WHICH TIME THE ALTITUDE WAS 30,000 METERS; THEN
WESTWARD [redacted] E AT 142 HRS WHEN THE ALTITUDE WAS 20,000 METERS;
TRACKING CONTINUED SOUTHWARD [redacted] E AT 1543 HRS WHEN THE ALTITUDE
WAS 18,000 METERS AND FINALLY TO THE LAST NOTED POSITION [redacted] E AT
1550 HRS. [redacted] UNIDENTIFIED FLYING OBJECT SPEED WAS
COMPUTED FROM CHART. 2. ALTHOUGH THE FLYING OBJECT WAS NOT IDENTIFIED, IT WAS
INFERRED TO HAVE BEEN A BALLOON ON THE BASIS OF ITS ALTITUDES AND OTHER
CONDITIONS.) CLASSIFIED BY RECORDED REPORTING OFFICER. EXEMPT FROM GENERAL
DECLASSIFICATION SCHEDULE OF E. O. 11652. EXEMPTION CATEGORY 5B(2). IMPOSSIBLE
TO DETERMINE DATE OF AUTOMATIC DECLASSIFICATION. 300 #0911

[redacted]

126

CATEGORY = 405 M = 63

MESSAGE = 02641439

XO #0792 2600750
ZNY MMORN
ZKZK RR [redacted] DE
[redacted]

ZEM
T O P S E C R E T UMBRA [redacted] (NOFORN [redacted] WARNING NOTICE: SENSITIVE INTELLIGENCE SOURCES AND METHODS INVOLVED) THE FOLLOWING INFORMATION HAS BEEN RECEIVED [redacted] RADAR TRACKING OF BALLOON AND UNIDENTIFIED OBJECT FLIGHTS IS DATED [redacted] THIRTEEN FLIGHTS OF (BALLOONS AND UNIDENTIFIED OBJECTS ALONG [redacted] ON [redacted] PLOTTED (BALLOON AND UNIDENTIFIED OBJECT POSITIONS ARE APPROXIMATE.) ON 12 [redacted] A TOTAL OF 13 FLIGHTS OF BALLOON AND UNIDENTIFIED FLYING OBJECTS WERE EITHER TRACKED OR SPOTTED AS FOLLOS:

AN UNIDENTIFIED FLYING OBJECT AT UNKNOWN ALTITUDE AND 60 KMH WAS TRACKED NORTHWARD FROM [redacted] AT 0813 HRS TO [redacted] AND THEN NORTHWESTWARD TO [redacted] AND FINALLY WESTWARD TO [redacted] AT 0924 HRS. 4. AN UNIDENTIFIED FLYING OBJECT AT UNKNOWN ALTITUDE AND 50 KMH WAS TRACKED NORTHWESTWARD FROM [redacted] E AT 0813 HRS TO [redacted] AND THEN WESTWARD TO [redacted] AT 0900 HRS.

6. AN UNIDENTIFIED FLYING OBJECT AT UNKNOWN ALTITUDE AND 30 WAS TRACKED WESTWARD FROM [redacted] E AT 0805 HRS TO [redacted] AT 1000 HRS. 7. AN UNIDENTIFIED FLYING OBJECT AT UNKNOWN ALTITUDE AND SPEED WAS TRACKED SOUTHWESTWARD FROM [redacted] AT 0929 HRS TO [redacted] AT 0946 HRS. 8. AN UNIDENTIFIED FLYING OBJECT AT UNKNOWN ALTITUDE AND SPEED WAS TRACKED WESTWARD

127

FROM [redacted] AT Ø93Ø HRS TO [redacted] AT Ø936 HRS. 9. AN UNIDENTIFIED FLYING OBJECT AT 22,ØØØ METERS ALTITUDE AND UNKNOWN SPEED WAS TRACKED NORTHWARD FROM [redacted] AT Ø941 HRS TO [redacted] AT Ø944 HRS.

1. BALLOON AN[redacted]UNIDENTIFIED FLYING OBJECT SPEEDS WERE COMPUTED FROM CHART. 2. IDENTIFICATION DATA FOR THE UNIDENTIFIED FLYING OBJECTS WERE NOT ABAILABLE AS THEIR TRACKING WA ALREADY IN PROGRESS WHEN [redacted]. HOWEVER, THEY WERE INFERRED TO HAVE BEEN BALLOONS IN LIGHT OF THE CONDITION OF THEIR APPARANCE AND THEIR INDICATED ALTITUDES.) CLASSIFIED BY RECORDED REPORTING OFFICER. EXEMPT FROM GENERAL DECLASSIFICATION SCHEDULE OF E. O. 11652. EXEMPTION CATEGORY 5B(2). IMPOSSIBLE TO DETERMINE DATE OF AUTOMATIC DECLASSIFICATION. 8ØØ #Ø792

[redacted]

#CATEGORY = 4Ø5 M = 65

MESSAGE = Ø2657171

XO #Ø892 282Ø938
ZNY MMCRN
ZKZK RR [redacted] DE
[redacted]
ZEM
[redacted] WARNING NOTICE: SENSITIVE INTELLIGENCE SOURCES AND METHODS INVOLVED. THE FOLLOWING INFORMATION HAS BEEN RECEIVED [redacted] RADAR TRACKING OF BALLOON AND UNIDENTIFIED OBJECT FLIGHTS I[redacted] TRACKING OF A BALLOON FLIGHT [redacted] AND AN UNIDENTIFIED OBJECT FLIGHT [redacted] PLOTTED BALLOON AND UNIDENTIFIED OBJECT POSITIONS ARE APPROXIMATE [redacted]

[redacted] 2. ON [redacted] AN UNIDENTIFIED FLYING OBJECT AT APPROXIMATELY 9Ø KMH WAS TRACKED IN [redacted] NORTHWEST OF [redacted] THE FLYING OBJECT WAS INITIALLY DETECTED AT [redacted] AT 9,5ØØ METERS ALTITUDE AT 1415 HRS AND TRACKED TO [redacted] AND THEN TO [redacted] AT 15Ø5 HRS AT WHICH TIME THE ALTITUDE WAS 24,5ØØ METERS; AT 1529 HRS IT WAS TRACKED TO [redacted], TO [redacted] AT 1537 HRS AT WHICH TIME THE ALTITUDE WAS

[redacted]

129

16,000 METEPS AND FINALLY TC THE LAST NCTED PCSITION OF [redacted] AT 1542 HRS AT 16,000 METERS ALTITUDE. [redacted] 1. SPEEKS FOR BOTH FLIGHTS WERE COMPUTED FRCM CHART. 2. THE UNIDENTIFIED FLYING OBJECT WAS INFERRED TC BE A BALLCCN ON THE BASIS OF ITS ALTITUDE, SPEED AND OTHER CONDITIONS.) CLASSIFIED BY RECORDED REPORTING CFFICER. EXEMPT FROM GENERAL DECLASSIFICATION SCHEDULE CF E.O. 11652. EXEMPTION CATEGORY 5B(2). IMPOSSIBLE TO DETERMINE DATE CF AUTOMATIC DECLASSIFICATICN. #0892

T O P S E C R E T U [redacted]

CATEGORY = 405 M = 64

MESSAGE = 02641441

XO #0794 2600850
ZNY MMORN
ZKZK RR [redacted] DE
[redacted]
ZEM
T O P S E C R E T SHOR [redacted] NOFORN [redacted] WARNING NOTICE: SENSITIVE INTELLIGENCE SOURCES AND METHODS INVOLVED. THE FOLLOWING INFORMATION HAS BEEN RECEIVED [redacted] RADAR TRACKING OF BALLOON AND UNIDENTI- FIED OBJECT FLIGHTS [redacted] TEN BALLOON AND UNIDENTIFIED OBJECT FLIGHTS [redacted] PLOTTED BALLOON AND UNIDENTIFIED OBJECT POSITIONS ARE APPROXIMATE. TEN BALLOON AND UNIDENTIFIED OBJECT FLIGHTS OVER [redacted] WERE EITHER TRACKED OR SPOTTED AS FOLLOWS: 1. A BALLOON FLIGHT AT UNKNOWN ALTITUDE (UNK ALT) AND 90 KMH WAS TRACKED WESTWARD FROM [redacted] AT 0450 HRS TO [redacted] AND NORTHWESTWARD TO [redacted] AT 0746 HRS.

8. AN UNIDENTIFIED FLYING OBJECT AT UNK ALT AND 130 KMH WAS TRACKED NORTHWESTWARD FROM [redacted] AT 0556 HRS TO [redacted] AND THEN WEST-SOUTHWESTWAD TO [redacted] AT 0746 HRS. 9.

T O P S E C R E T U M B R A

131

10. AN
UNIDENTIFIED OBJECT AT UNK ALT AND SPEED WAS SPOTTED AT [redacted] AT 0744 HRS. ([redacted] 1. BALLOON AND UNIDENTIFIED OBJECT SPEEDS WERE COMPUTED FROM CHART. 2. ALTHOUGH TARGET IDENTIFIECATION DATA FOR THE UNIDENTIFIED OJBECTS OF PARAGRAPHS 8 AND 10 ARE UNAVAILABLE, THEY WERE INFERRED TO HAVE BEEN BALLOONS ON THE BASIS OF THEIR APPEARANCE, INDICATED ALTITUDES, ETC.) CLASSIFIED BY RECORDED REPORTING OFFICER. EXEMPT FROM GENERAL DECLASSIFIED BY RECORDED REPORTING OFFICER. EXEMPT FROM GENERAL DECLASSIFICATION SCHEDULE OF E.O. 11652. EXEMPTION CATEGORY 58(2). IMPOSSIBLE TO DETERMINE DATE OF AUTOMATIC DECLASSIFICATION. 700 #0794

~~T O P S E C R E T U M B R A~~

= 4Ø5 M = 68

= Ø2668526

2981818

H

[redacted] DE

[redacted]

(~~S E C R E T U M B R A~~)

[redacted] 3/CC/35732-74 [redacted]

UNIDENTIFIED FLYING OBJECT

[redacted]

OUS OBJECT WAS SIGHTED [redacted]

[redacted] AND AFTER SEVERAL MINUTES IT DISAPPEARED FROM VIEW. [redacted]

~~T O P S E C R E T U M B R A~~

133

CATEGORY = 4Ø5 M = 3

MESSAGE = Ø2741Ø91

XO #ØØ71 Ø28Ø815
ZNY MMCRN
ZKZK RR [redacted] DE
[redacted]

ZEM
~~(S E C R E T SPOKE~~ [redacted] " [redacted] WARNING NOTICE: SENSITIVE INTELLIGENCE SOURCES AND METHODS INVOLVED) (THE FOLLOWING INFORMATION HAS BEEN RECEIVED) [redacted] RADAR TRACKING [redacted] A FLYING OBJECT [redacted] FLIGHT OF AN UNIDENTIFIED OBJECT [redacted] PLOTTED POSITIONS OF THE UNIDENTIFIED FLYING OBJECT ARE APPROXIMATE) [redacted] AN UNIDENTIFIED OBJECT FLYING (AT ALTITUDES OF 6,ØØØ THEN 24,ØØØ AND FINALLY 4,ØØØ METERS AND 14Ø KMH) WAS INITIALLY DETECTED [redacted] THE UNIDENTI- FIED OBJECT'S FLIGHT WEST OF [redacted]

THE REMAINDER OF THE ENTIRE FLIGHT WAS IN [redacted] 1. ALTHOUGH TARGET IDENTIFICATION DATA WAS NOT AVAILABLE, IT WAS INFERRED TO HAVE BEEN A BALLOON ON THE BASIS OF ITS ALTITUDE, SPEED AND OTHER CONDITIONS. 2. THE SPPEED OF THE UNIDENTIFIED OBJECT WAS COMPUTED FROM THE CHART.) (CLASSIFIED BY RECORDED REPORTING OFFICER. EXEMPT FROM GENERAL DECLASSIFICATION SCHEDULE OF E.O. 11652. EXEMPTION CATEGORY 5B(2). IMPOSSIBLE TO DETERMINE DATE OF AUTOMATIC DECLASSIFICATION. 3ØØ #ØØ71)

CATEGORY = 405 M = 20

MESSAGE = 03201877

LA #0869 3182301
ZNY MMNSH
ZKZK RR [redacted] DE
[redacted]
ZEM
[redacted]
XXMM
[redacted] 2/[redacted]/5365-76 [redacted] REPORTS SIGHTING OF UNIDENTIFIED FLYING OBJECT
XXCC

[redacted] REPORTED THE SIGHTING OF AN UNIDENTIFIED FLYING OBJECT ON [redacted] IN A REPORT [redacted] THAT A "LUMINOUS OBJECT" OVERFLEW [redacted] AND WAS HEADING IN THE DIRECTION OF [redacted]. XGDS 2 XXHH #0869 NNN

[redacted]

CATEGORY = 405 M = 18

MESSAGE = 03114155

LA #0072 1751421
ZNY MMNSH
ZKZK RR [redacted] DE
[redacted]
ZEM
S E C R E T SPOKE
XXMM
[redacted] 2/[redacted]/2899-76 [redacted] REPORT SIGHTING 1
OF UNIDENTIFIED FLYING OBJECT
XXCC

AN UNIDENTIFIED FLYING OBJECT WAS SIGHTED BY [redacted] BETWEEN THE [redacted] REGIONS [redacted] BETWEEN 2130Z AND 2140Z. THE OBJECT WAS DESCRIBED AS HAVING A SEMI-CIRCLE SHAPE AND LOOKED LIKE AN ARC IN THE SKY. THIS OBJECT WAS ALSO NOTED AS BEING WHITE AND VERY LARGE. IT WAS SEEN FOR A PERIOD OF ABOUT TEN MINUTES AND IT SEEMED TO JUST HANG IN THE SKY FOR A FEW MINUTES BEFORE MOVING ON IN A WESTERLY DIRECTION. XGDS 2 XXHH #0072 NNN

TOP SECRET UMBRA

136

~~TOP SECRET UMBRA~~

CATEGORY = 405 M = 15

MESSAGE = 02828303

AH #8581 1502023
ZNY MMNSH
ZKZK RR [redacted] DE
[redacted]
FM DIRNSA
[redacted]
ZEM
~~TOP SECRET UMBRA~~
XXMM
[redacted] 3/[redacted]/17412-75 [redacted] UNIDENTIFIED FLYING OBJECT
XXCC

[redacted] A REPORT [redacted] STATES THAT [redacted] A SMOKING LUMINOUS OBJECT MOVED THROUGH THE AIR AT AN ALTITUDE OF APPROXIMATELY 600 METERS FROM [redacted] TOWARDS [redacted] AT THE BORDER OPPOSITE [redacted] THE SMOKE RADIUS EXTENDED ABOUT 100 METERS FROM THE CENTER OF THE LUMINOUS OBJECT AND A GREEN LIGHT COULD BE SEEN FOLLOWING IT. THE SMOKE RADIUS GRADUALLY INCREASED IN SIZE AND DISAPPEARED. THE LUMINOUS OBJECT WHICH WAS IN THE CENTER OF THE SMOKE ALSO DISAPPEARED. [redacted]
[redacted] XGDS 2 XXHF #8581 NNN

~~TOP SECRET UMBRA~~

137

[redacted]

CATEGORY = 4?? M = 40

MESSAGE = #3626469

LH #2220 134/546
ZNY MMNSH
ZKZK PP [redacted] DE
[redacted]
ZEM
[redacted] INTEL FINAL SECTION OF SEVEN [redacted]
THE FOLLOWING INFORMATION HAS BEEN PROVIDED [redacted]

F. RADAR TRACKING OF [UNIDENTIFIED FLYING OBJECTS: 16.]

T[redacted]

138

[redacted]

CATEGORY = 140F M = 47

MESSAGE = 03626467

LH #0013 1810549
ZNY MMNSH
ZKZK PP [redacted] DE
[redacted]
ZEM
[redacted] SECTION NINE [redacted] THE
FOLLOWING INFORMATION HAS BEEN PROVIDED [redacted]

T [redacted]

139

8. RADAR TRACKING OF UNIDENTIFIED FLYING OBJECTS (UFO): A. Ø2Ø6-Ø445, FIVE UFO (PROBABLY BALLOONS) MOVED SLOWLY FROM THE VICINITY OF [redacted] WESTWARD. B. Ø73Ø-Ø94Ø, FOUR UFO (PROBABLY BALLOONS) MOVED SLOWLY FROM WEST OF [redacted] TOWARD WEST, ALT 4Ø,2ØØ- 62,4ØØ. C. Ø645-152Ø, TEN UFO (PROBABLY BALLOONS) MOVED FROM 8Ø EAST OF [redacted] PASSED [redacted], [redacted] AND [redacted] AND FADED [redacted], ALT 6Ø,ØØØ-82,ØØØ FT.

[redacted]

CATEGORY = 4Ø5 M = 46

MESSAGE = Ø3626446

LH #ØØ15 1930625
ZNY MMNSH
ZKZK PP [redacted] DE
[redacted]
ZEM
[redacted] SECTION [redacted] THE FOLLOWIN
INFORMATION HAS BEEN PROVIDED [redacted]

6. RADAR TRACKING OF UFO ACTIVITIES: A. Ø52Ø-Ø544, ONE UFO (BELIEVED BALLOON) MOVED OVER [redacted] EAST OF [redacted]. B. Ø712-Ø81Ø, SIX UFOS (BELIEVED BALLOONS) MOVED WESTWARD OVER NORTH OF [redacted] AND [redacted] AREAS OF [redacted] ALT 59,Ø3Ø-62,4ØØ FT.

[redacted]

141

[redacted]

CATEGORY = 405 Y = 45

MESSAGE = Ø3626432

LH #Ø012 179Ø633
ZNY MMNSH
ZKZK PP [redacted] DE
[redacted]
ZEM
[redacted] SECTION 9 [redacted] THE FOLLOWIN
INFORMATION HAS BEEN PROVIDED [redacted]

5. RADAR TRACKING OF UNIDENTIFIED FLYING OBJECTS (UFO): Ø429-ØØ26 SEVEN UFO (PROBABLY BALLOONS) MOVED SLOWLY FROM WEST OF [redacted] TOWARD SW, PASSED SOUTH OF [redacted] AND [redacted] ALT 32,8ØØ-66,ØØØ FT.

[redacted]

142

[redacted]

CATEGORY = 4Ø5 N = 44

MESSAGE = 33626417

LH #ØØØ8 1783441
ZNY MMNSH
ZKZK PP [redacted] DE
[redacted]
ZFM ;
[redacted] SECTION SIX [redacted] THE
FOLLOWING INFORMATION HAS BEEN PROVIDED [redacted]

6. [redacted] TO <u>UNIDENTIFIED FLYING OBJECTS</u> (UFO): Ø740-1Ø34, [redacted] RADAR TRACKED SEVEN UFO (PROBABLY BALLOONS) MOVED FROM NW OF [redacted] TOWARD WEST, PASSED [redacted] AND [redacted] ALT 58,ØØØ-72,1ØØ FT.

[redacted]

143

~~T O P S E C R E T U M B R A~~

CATEGORY = 405 M = 43

MESSAGE = 03636403

L4 #0007 1770737
ZNY MMNSH
ZKZK PP [redacted] DE
[redacted]
ZEM
~~S E C R E T~~ SPOKE [redacted] UNINTEL SECTION FOUR [redacted] THE
FOLLOWING INFORMATION HAS BEEN PROVIDED [redacted]

5. RADAR TRACKING OF
<u>UNIDENTIFIED FLYING OBJECTS</u> (UFO): 0522-0746, TWO UFO (PROBABLY BALLOONS)
MOVED SLOWLY FROM SOUTH OF [redacted] TOWARD WEST, ALT
32,8000-46,000 FT.

~~T O P S E C R E T U M B R A~~

144

[redacted]

CATEGORY = 405 M = 42

MESSAGE = 03626397

LH #0311 1740536
ZNY MMNSH
ZKZH PP [redacted] DE
[redacted]

ZEM .
[redacted] SECTION 6 [redacted] THE FOLLOWING
INFORMATION HAS BEEN PROVIDED [redacted]

6. RADAR TRACKING OF

[redacted]

145

<u>UNIDENTIFIED FLYING OBJECTS</u> (UFO): A. 0413-1414, SIX UFO (PROBABLY BALLOONS) MOVED FROM SW OF [redacted] TOWARD WEST AND PASSED [redacted] AND [redacted] ALT 15,500-62,400 FT. B. 1441-1606, ONE UFO (PROBABLY A BALLOON) MOVED FROM 49 NAUTICAL MILES 360 DEGREES OFF [redacted] TOWARD WEST, PASSED [redacted] AND FADED [redacted], ALT 85,300 FT.

146

[redacted]

~~TOP SECRET UMBRA~~

CATEGORY = 4?5 M = 41

MESSAGE = Ø3626387

LH #ØØ13 173Ø628
ZNY MMNSH
ZKZK PP [redacted] DF
[redacted]
ZEM
S E C R [redacted] SECTION 7 [redacted] THE FOLLOWING
INFORMATION HAS BEEN PROVIDED [redacted]

6. RADAR TRACKING OF UNIDENTIFIED FLYING OBJECTS
(UFO)(C) A. Ø3Ø5-123Ø, 22 UFO (PROBABLY BALLOONS) MOVED FROM WEST OF HSINCHOU
TOWARD SW, PASSED [redacted] AND [redacted], ALT 54,ØØØ-96,ØØØ FT. B. Ø919-16Ø7, 21 UFO (PROBABLY BALLOONS) MOVED
OVER [redacted] AND
[redacted] ALT 82,ØØØ-96,ØØØ FT.

~~TOP SECRET~~

147

, [redacted] RADAR TRACKED UNKNOWN FLYING OBJECTS (PROBABLY BALLOONS) IN 24 FLIGHTS FROM THE VICINITY OF [redacted] MOVING SLOWLY OVER [redacted] ALT 17,000-28,000 METERS.

148

T O P [redacted]

CATEGORY = 4Ø5 M = 4Ø.

MESSAGE = Ø3626376

LH #ØØ12 172Ø621
ZNY MMNSH
ZKZK PP [redacted] DE
[redacted]

ZEM
[redacted] SECTION 6 [redacted] THE FOLLOWING
INFORMATION HAS BEEN PROVIDED: [redacted]

T O P [redacted]

149

5. [redacted]

REACTION TO UFO (BELIEVED BALLOONS) ACTIVITIES: 0400-1205, [redacted] RADAR TRACKED EIGHT UFOS MOVING WESTWARD OVER WEST OF [redacted] AND [redacted], ALT 43,600-79,000 FT.

~~T O P S E C R E T U M B R A~~

4Ø5 M = 29

Ø3626264

71Ø61Ø

[redacted] DE

[redacted] L SECTION 7 [redacted] THE FOLLOWING
ON HAS BEEN PROVIDED [redacted]

5. RADAR TRACKING OF UNIDENTIFIED FLYING OBJECTS (UFO): A.
[illegible], TEN UFO MOVED FROM NW OF MAKUNG, PASSED [redacted] AND [redacted]
TOWARD WEST, ALT 69,ØØØ-79,ØØØ FT. B. Ø251-Ø546, SEVEN UFO MOVED FROM
[redacted], PASSED [redacted] AND [redacted] TOWARD WEST, ALT
5,3ØØ FT. C. Ø752-Ø9Ø7, NINE UFO MOVED RESPECTIVELY FROM [redacted]

~~T O P S E C R E T U M B R A~~

151

[redacted] AND [redacted] TOWARD WEST, ALT 55,333 FT.

[redacted]

CATEGORY = 485 M = 37

MESSAGE = 03626213

LN #0317/2 1742717
ZNY MMCRN
ZKZK RR [redacted] DE
[redacted]
ZEM
[redacted] WARNING NOTICE-SENSITIVE INTELLIGENCE SOURCES AND METHODS INVOLVED THE FOLLOWING INFORMATION HAS BEEN RECEIVED [redacted]

T O P S E C R E T [redacted]

153

10. ON 5 JUNE TWO UNIDENTIFIED FLYING OBJECTS FLEW OVER THE [redacted] NORTH OF [redacted] FROM 1142 THROUGH 1148. THE OBJECTS WERE INDICATED AS FRIENDLY AND NUMBERING TWO BUT THAT THEY WERE UN- IDENTIFIABLE. THEIR ALTITUDE WAS NOT INDICATED. THEIR SPEED AS COMPUTED ON CHART WAS APPROXIMATELY 930 KMH. WHETHER THEY WERE AIRCRAFTS, MISSILES OR SOME OTHER FLYING OBJECTS COULD NOT BE DETERMINED.

CATEGORY = 405 M = 34

MESSAGE = 33615850

LH +0005 1700006
ZNY MMNSH
ZKZK PP [redacted] DE
[redacted]
ZEM
SECRET SPOKE [redacted] SEC 5 CITE [redacted] THE FOLLOWING
INFORMATION HAS BEEN PROVIDED [redacted]

155

[redacted]

7. [redacted] REACTION TO <u>UNIDENTIFIED FLYING OBJECTS</u> (UFO): 0527-1230, [redacted] RADAR TRACKED EIGHT UFO (PROBABLY BALLOONS) MOVING FROM NW OF [redacted] TOWARD WEST AND PASSING [redacted] AND [redacted] ALT 43,600-72,100 FT.

[redacted]

156

T O P S E C R E T U [redacted]

CATEGORY = 405 M = 35

MESSAGE = 03615382

LH #0011 1700806
ZNY MMNSH
ZKZK PP [redacted] DE
[redacted]
ZEM
S E C R E T SPOKE [redacted] SEC 7 [redacted] THE FOLLOWING
INFORMATION HAS BEEN PROVIDED: [redacted]

F. RADAR TRACKING OF <u>UNIDENTIFIED FLYING OBJECTS</u>: FOUR.

T O P S E C R E T U [redacted]

157

TOP SECRET U[redacted]

CATEGORY = 405 Y = 33

MESSAGE = Ø361523Ø

LH #ØØ13 1670536
ZNY MMNSH
ZKZK PP [redacted] DE
[redacted]
ZEM
S [redacted] WNINTEL SECTION EIGHT [redacted] THE
FOLLOWING INFORMATION HAS BEEN PROVIDED: [redacted]

[redacted]

158

9. RADAR TRACKING OF UNIDENTIFIED FLYING OBJECTS (UFO): A. 0333-0527, ONE UFO (PROBABLY A BALLOON) MOVED FROM 51 NM NW OF [redacted] PASSED NORTH OF [redacted] AND NORTH OF [redacted] TOWARD WEST, ALT 66,000-9,000 FT. B. 0935-0957, FOUR UFO (PROBABLY BALLOONS) MOVED OVER [redacted] AND [redacted], ALT 66,000-72,100 FT.

~~TOP SECRET UMBRA~~

CATEGORY = 485 Y = 82

MESSAGE = 03615926

LH #0016 1662701
ZNY MMNSH
ZKZK PP [redacted] DE
[redacted]
ZEM
~~TOP SECRET SPOKE~~ [redacted] ~~NINTEL~~ SECTION 7 [redacted] THE FOLLOWING
INFORMATION HAS BEEN PROVIDED: [redacted]

7. RADAR TRACKING OF
UNIDENTIFIED FLYING OBJECTS: A. 0521-0835, FOUR UFO (PROBABLY BALLOONS) MOVED
OM WEST OF [redacted] PASSED OVER [redacted] AND [redacted]. ALT 82,000- 85300 FEET.
1440-1442, ONE UFO (PROBABLY A BALLOON) MOVED OVER [redacted], ALT 82,000-85,300
FT. C. [redacted] 0430-1100, NINE UFO'S MOVED WESTWARD OVER [redacted] AND
[redacted], ALT 62,400-82,000 FEET. (DELAYED INFORMATION)

~~TOP SECRET UMBRA~~

160

[redacted]

[redacted]

CATEGORY = 405 M = 3Ø

MESSAGE = Ø261575F

LH #Ø014 1634717
ZNY MMNSH
ZKZK PP [redacted] DE
[redacted]

ZEM
S E C R E T WPUKE [redacted] ZNINTEL SECTION 5 [redacted] THE FOLLOWING
INFORMATION HAS BEEN PROVIDED [redacted]

[redacted]

161

5. UFO ACTIVITIES: Ø224-Ø8Ø3, SIX UFOS MOVED OVER NW OF [redacted], AND [redacted] AREA, ALT 62,4ØØ-69,ØØØØ FT.

162

[redacted]

CATEGORY = 403 M = 31

MESSAGE = 23615787

LH #0016 1630717
ZNY MMNSH
ZK7K PP [redacted] DE
[redacted]
ZEM
S E C R E T S UNI [redacted] WNINTEL SECTION 7 [redacted] THE FOLLOWING
INFORMATION HAS BEEN PROVIDED [redacted]

E. RADAR TRACKING OF UNIDENTIFIED FLYING OBJECTS: SEVEN.

T [redacted] U M B R A

163

T O P S E C R E T U M B R A

TEGORY = 405 M = 29

SSAGE = 73615741

#0012 1570616

Y MMNSH

ZK PP [redacted],DE

[redacted]

EM.

S E C R E T SPOKE [redacted] SECTION 8 [redacted] THE FOLLOWING

NFORMATION HAS BEEN PROVIDED [redacted]

6. RADAR TRACKING OF UNIDENTIFIED FLYING OBJECTS (UFO): I. 1023-1133 AND

T O P S E C R E T U M B R A

164

[redacted]

1353-1651, TWO UFO MOVED RESPECTIVELY FROM NW OF [redacted] TOWARD WEST, ALT 66,000- 75,500 FT. 2. 1421-1555, ONE UFO MOVED FROM NW OF [redacted] TOWARD WEST AND PASSED [redacted]

[redacted]

165

[redacted]

CATEGORY = 405 M = 27

MESSAGE = 33615728

LH #0028 1560718
ZNY MMNSH
ZKZK PP [redacted] DE
[redacted]
ZEM
[redacted] SECTION 6 [redacted] THE FOLLOWING
INFORMATION HAS BEEN PROVIDED [redacted]

~~TOP SECRET UMBRA~~

166

15. RADAR TRACKING OF UNIDENTIFIED FLYING OBJECTS (UFO): Ø12Ø-135Ø, NINE UFO (PROBABLY BALLOONS) MOVED FROM SOUTH OF [redacted] CROSSED [redacted] AND PASSED OVER [redacted] AND [redacted], ALT 6,6ØØ-76,2ØØ FT.

167

[redacted]

CATEGORY = 425 M = 28

MESSAGE = 33615734

LH #0030 1563719
ZNY MMNSH
ZKZK PP [redacted] DE
[redacted]
ZEN
[redacted] SECTION 8 [redacted] THE FOLLOWING
INFORMATION HAS BEEN PROVIDED [redacted]

F. RADAR TRACKING OF UNIDENTIFIED FLYING OBJECTS: ONE.

~~TOP SECRET UMBRA~~

168

CATEGORY = 405 M = 25

MESSAGE = Ø3604633

LH #ØØ11 1537516
ZNY MMNSH
ZKZK PP [redacted] DE
[redacted]
ZEM.
[redacted] E [redacted] ~~UNINTEL~~ FINAL SECTION OF SEVEN [redacted] THE FOLLOWING INFORMATION HAS BEEN PROVIDED [redacted]

5. <u>UFO</u> ACTIVITIES: (ALT IN FT) A. ØØ15-ØØØØ, EIGHT UFOS MOVED OVER THE [redacted], AND [redacted] ALT 56,ØØØ-99,ØØØ. B. Ø315-127Ø, FOUR UFOS MOVED OVER THE [redacted] AND [redacted], ALT 56,ØØØ-96,ØØØ. C. Ø424-1257, FIVE UFOS MOVED OVER [redacted], ALT (STET) 56,ØØØ-93,ØØØ. D. Ø638-1257, SIX UFOS MOVED OVER [redacted], AND [redacted] ALT 56,ØØØ-755ØØ. E. 1653-182Ø, ONE UFO MOVED OVER AREA 11 NAUTICAL MILES SW OF [redacted] AND MOVED WESTWARD OVER 2Ø NAUTICAL MILES WEST OF [redacted], ALT 62,2ØØ.

169

T O P S [redacted]

= 405 M = 24

03604622

1520747
ZZS
[redacted] DE

[redacted] UNINTEL SECTION SIX [redacted]
INFORMATION HAS BEEN PROVIDED [redacted] THE
[redacted]

T O P [redacted]

170

[redacted].5. RADAR TRACKING OF UNIDENTIFIED FLYING OBJECTS (UFO): Ø215-162Ø, TEN UFO (PROBABLY BALLOONS) MOVED FROM WEST OF [redacted] CROSSED THE STRAIT AND PASSED [redacted] AND [redacted] ALT 33,ØØØ- 99,ØØØ FT.

[redacted]

05 M = 21

04599

641

[redacted] DE

[redacted]

S [redacted] SECTION SIX [redacted] THE
ORMATION HAS BEEN PROVIDED [redacted]

6. RADAR TRACKING OF UNIDENTIFIED FLYING OBJECTS (UFO):
E UFO MOVED OVER [redacted] ALT 15,000-30,000 FT.

[redacted]

172

T [redacted]

ATEGORY = 4Ø5 M = 22

ESSAGE = Ø36Ø46Ø1

#ØØ32 15ØØ641
Y MMNSH
ZK PP [redacted] DE
[redacted]

M
[redacted] SECTION EIGHT [redacted] THE
LLOWING INFORMATION HAS BEEN PROVIDED BY [redacted]

T [redacted]

173

[redacted]

[redacted]

OF UNIDENTIFIED FLYING OBJECTS: NINE.

[redacted]

174

[redacted]

CATEGORY = 405 M = 23

MESSAGE = 036Ø46Ø2

LH #ØØ33 15ØØ641
ZNY MMNSH
ZKZK PP [redacted] DE
[redacted]
ZEM
S[redacted] , [redacted] SECTION NINE [redacted] THE
FOLLOWING INFORMATION HAS BEEN PROVIDED [redacted]

[redacted]

TOP SECRET UMBRA

ED FLYING OBJECTS: 22. E. RADAR TRACKING OF

T O P S E C R E T U M B R A

176

T O P S E C R E T U [redacted]

CATEGORY = 4Ø5 M = 2Ø

MESSAGE = Ø36Ø4589

LH #ØØ12 146Ø716
ZNY MMNSH
ZKZK PP [redacted] DE
[redacted]

ZEM
S E C R E T [redacted] SEC 9 CITE [redacted] THE FOLLOWING
INFORMATION HAS BEEN PROVIDED [redacted]

T O P [redacted] U M B R A

177

7. RADAR TRACKING OF UNIDENTIFIED FLYING OBJECTS (UFO): 0719-1453, NINE UFO (PROBABLY BALLOONS) MOVED FROM WEST OF [redacted], CROSSED [redacted], ENTERED [redacted] AND PASSED [redacted] AND [redacted] ALT 35,000-100,000 FT.

178

~~TOP SECRET U~~

CATEGORY = 405 M = 10

MESSAGE = 03604576

LH #0014 1450550
ZNY MMNSH
ZKZK PP[redacted] DE
[redacted]
ZEM
S E C R E T [redacted] SEC 8 [redacted] THE FOLLOWING
INFORMATION HAS BEEN PROVIDED: [redacted]

~~TOP SECRET UMBRA~~

179

6. RADAR TRACKING OF UNIDENTIFIED FLYING OBJECTS (UFO): A. Ø555-132Ø, THREE UFO MOVED FROM THE [redacted] NEAR [redacted] CROSSED [redacted] AND FADED NEAR [redacted] ALT 69,ØØØ-82,ØØØ FT. B. Ø747-Ø91Ø, TWO UFO MOVED FROM 17 NM EAST OF [redacted] PASSED [redacted] TOWARD [redacted], ALT 82,ØØØ- 85,ØØØ FT. C. Ø922-131Ø, ONE UFO MOVED FROM 47 NM NW OF [redacted], PASSED SOUTH OF [redacted] AND FADED NEAR [redacted] ALT 72,1ØØ FT. D. 11Ø8-132Ø, ONE UFO MOVED FROM 45 NM NW OF [redacted], PASSED SOUTH OF [redacted] AND FADED [redacted] ALT 65,ØØØ FT.)

T[redacted]

CATEGORY = 485 M = 18

MESSAGE = 33604555

LH #0015 1438814
ZNY MMNSH
ZK7K PP [redacted] DE
[redacted]
ZEM
S E C R E T [redacted] SECTION FIVE [redacted] THE
FOLLOWING INFORMATION HAS BEEN PROVIDED [redacted]

6. RADAR TRACKING OF

TOP SECRET UMBRA

181

UNIDENTIFIED FLYING OBJECTS (UFO): 0235-0730, THREE UFO MOVED FROM [redacted], PASSED [redacted] AND NE OF [redacted] TOWARD NW, ALT 75,500-85,300 FT.

~~TOP SECRET UMBRA~~

CATEGORY = 405 M = 17

MESSAGE = 73594581

LH #0010 1370627
ZNY MMNSH
ZKZK PP [redacted] DE
[redacted]

ZEM
~~SECRET GROUP~~ [redacted] ~~WNINTEL~~ SEC 7 [redacted] THE FOLLOWING INFORMATION HAS BEEN PROVIDED [redacted]

7. UFO ACTIVITIES: (ALT IN FT) A. 0744-0940, THREE UFOS MOVED NORTHWESTWARD OVER NE OF [redacted] AND [redacted] ALT 20,000-86,800. B. 0756-0923, ONE UFO MOVED NORTHWESTWARD OVER NE OF [redacted]G, AND NW OF [redacted] ALT 85,300. C. 0906-0935, FOUR UFOS MOVED NORTHWESTWARD OVER NW OF [redacted] AND [redacted] ALT 20,000-86,800. D. 0848-1405, ONE UFO MOVED WESTWARD OVER EAST OF [redacted] SOUTH OF [redacted]

T[redacted]ECRET UMBRA

183

[redacted] AND [redacted] ALT 3Ø,ØØØ- 75,5ØØ. F.
Ø84Ø-1544, ONE UFO MOVED WESTWARD OVER EAST OF [redacted] SOUTH OF [redacted] THEN
MOVED SOUTHEASTWARD, AND FADED AWAY AT SE OF [redacted] ALT 3Ø,ØØØ-72,1ØØ.

[redacted]

CATEGORY = 4Ø5 M = 15

MESSAGE = Ø3594560

LH #ØØ18 1353728
ZNY MMNSH
ZKZK PP[redacted] DE
[redacted]

ZEM
[redacted] L SECTION 3 [redacted] THE FOLLOWING
INFORMATION HAS BEEN PROVIDED [redacted]

5. RADAR TRACKING OF UNIDENTIFIED FLYING OBJECTS (UFO): A. Ø35Ø-Ø526. ONE UFO MOVED FROM 19 NAUTICAL MILES (NM) 18Ø DEGREES (DEG) OFF [redacted] TOWARD WEST AND FADED 27 NM EAST OF [redacted] ALT 25,ØØØ. B. Ø5Ø5-Ø52Ø, ONE UFO MOVED FROM 27 NM 18Ø DEG OFF [redacted] TOWARD NW AND FADED 3Ø NM EAST [redacted] C.

T [redacted]

185

0659-0700, ONE UFO MOVED 40 NM EAST OF [redacted] AND THEN FADED, ALT 8,000. D. 1005-1007, TWO UFO MOVED FROM 27 NM EAST OF [redacted] TOWARD EAST AND THEN FADED. E. 1001-1041, TWO UFO MOVED FROM 27 NM 90 DEG OFF [redacted] TOWARD SE AND FADED 37 NM EAST OF [redacted]

CATEGORY = 405 M = 16

MESSAGE = 23594561

LH #0019 1353723
ZNY MMNSH
ZKZK PP [redacted] DE
[redacted]
ZEM
~~S E C R E T SPOKE~~ [redacted] ~~WNINTEL~~ SECTION 4 [redacted] THE FOLLOWING
INFORMATION HAS BEEN PROVIDED [redacted]

187

[redacted]

E. RADAR TRACKING OF UNIDENTIFIED FLYING OBJECTS: FIVE.

T O P [redacted]

188

CATEGORY = 405 M = 14

MESSAGE = 03504554

LH #0017 1320723
ZNY MMNSH
ZKZK PP [redacted] DE
[redacted]

ZEM
[redacted] WNINTEL SECTION 10 [redacted] THE
FOLLOWING INFORMATION HAS BEEN PROVIDED [redacted]

189

5. RADAR TRACKING OF
LYING OBJECTS (UFO): 0247-0322 AND 0337-0322, THE UFO (PROBABLY
D OVER THE [redacted] 27 NAUTICAL MILES (NM) SE OF [redacted]
000-11,500 FT.

T O P S E C [redacted]

CATEGORY = 405 N = 13

MESSAGE = 03F04539

LH #0013 1310715
ZNY MMNSH
ZKZK PP [redacted] DE
[redacted]
ZEM
S E C R E T [redacted] FINAL SECTION 2 [redacted] THE FOLLOWING INFORMATION HAS BEEN PROVIDED [redacted]

8. RADAR TRACKING OF UNIDENTIFIED FLYING OBJECTS (UFO: A. Ø710-Ø800, ONE UFO MOVED SLOWLY SEVEN TO 17 NAUTICAL MILES SOUTH OF [redacted] ALT 6Ø,ØØØ- 65,ØØØ FT. B. Ø854-Ø9Ø3, ONE UFO (A BALLOON) MOVED OVER [redacted] ALT 41,4ØØ FT.

T O P [redacted]

191

[redacted]

CATEGORY = 405　　　　　　　　　　　　　　　M = 11

MESSAGE = 03594525

LH #0014 1300715
ZNY MMNSH
ZKZK PP [redacted] DE
[redacted]

ZEM
[redacted] SEC [redacted] THE FOLLOWING
INFORMATION HAS BEEN PROVIDED [redacted]

[redacted] SECRET UMBRA

192

, 7. UFO (BELIEVED BALLOONS)

ACTIVITIES: A. 0040-0732, ONE UFO MOVED OVER SE OF [redacted] EAST OF [redacted], AND [redacted], ALT 70,000 FT. B. 0415-0638, ONE UFO MOVED [redacted], AND EAST OF [redacted], ALT 24,200 FT. C. 0709-1157, ONE UFO MOVED [redacted] EAST OF [redacted]. D. 1015-1347, ONE UFO MOVED [redacted] #0014 NNN

193

T O P S E C R E T U

CATEGORY = 495 M = 12

MESSAGE = 03594526

LH #0015 1300715
ZNY MMNSH
ZKZK PP [redacted] DE
[redacted]

ZEM
[redacted] SEC 9 FINAL [redacted] THE
FOLLOWING INFORMATION HAS BEEN PROVIDED [redacted]

E. 1012-1045, ONE UFO MOVED OVER 25 NM 2?? DEGREES OFF [redacted]
11549), ALT 69,000 FT.

T O P S E C R E T U M B R A

194

T [redacted]

CATEGORY = 405 M = 17

MESSAGE = 03584703

LH #0013 1240626
ZNY MMNSH
ZKZK PP [redacted] DE
[redacted]
ZEM
[redacted] L SECTION 6 [redacted] THE FOLLOWING
INFORMATION HAS BEEN PROVIDED [redacted]

T O P [redacted] U M B R A

195

~~TOP SECRET U~~

5. RADAR TRACKING OF UNIDENTIFIED FLYING OBJECTS(UFO): A. 0938-0942, ONE UFO (PROBABLY A BALLOON) MOVED FROM 67 NAUTICAL MILES (NM) 262 DEGREES (DEG) OFF [redacted] AND FADED 59 NM 263 DEG OFF [redacted]. B. 0942-0944, ONE UFO (PROBABLY A BALLOON) MOVED FROM 38 NM 270 DEG OFF [redacted] AND FADED 86 NM 270 DEG OFF [redacted], ALT 79,000 FT.

~~TOP SECRET UMBRA~~

196

CATEGORY = 405 M = 6

MESSAGE = 03552308

LH #0013 0790647
ZNY MMNSH
ZKZK PP [redacted] DE
[redacted]
ZEM
SECRET [redacted] SECTION 5 [redacted] THE FOLLOWING
INFORMATION HAS BEEN PROVIDED [redacted]

197

5\.

RADAR TRACKING OF UFO ACTIVITIES: 1514-1515, ONE UFO WAS TRACKED OPERATING OVER SOUTH OF [redacted] #0013 NNN

[redacted]

CATEGORY = 405 M = ?

MESSAGE = 03553309

LH #0026 0670719
ZNY MMNSH
ZKZK PP [redacted] DE
[redacted]
ZEM
[redacted] L SECTION 6 [redacted] THE FOLLOWING
INFORMATION HAS BEEN PROVIDED [redacted]

5. RADAR TRACKING OF UFO ACTIVITIES: 1322-1342, ONE UFO (UNIDENTIFIED FLYING OBJECT) WAS TRACKED MOVING OVER AREA 43 NAUTICAL MILES 270 DEGREES OFF [redacted], MOVING TOWARD NORTHEAST, AND THEN FADING AWAY AT 21 NAUTICAL MILES ...DEGREES OFF [redacted], SPEED 51 KNOTS.

T [redacted] U M B R A

199

~~TOP SECRET UMBRA~~

CATEGORY = 405   N = 4

MESSAGE = 03529791

LH #0010 0417924
ZNY MMNSH
ZKZK PP [redacted] DE
[redacted]
ZEM
~~SECRET SPOKE~~ [redacted] ~~WNINTEL~~ [redacted] THE FOLLOWING
INFORMATION HAS BEEN PROVIDED [redacted]

~~TOP SECRET UMBRA~~

200

6. RADAR TRACKING OF UFO ACTIVITIES: 1446-1455, ONE UFO (BELIEVED BALLOON) WAS TRACKED MOVING NORTHEASTWARDS OVER AREA 40 NAUTICAL MILES 135 DEGREES OFF [redacted] AND FADING AWAY AT 43 NAUTICAL MILES 120 DEGREES OFF [redacted] ALT 46,000 FT.

CATEGORY = 405 N = 3

MESSAGE = 03511100

LH #0010 0300815
ZNY MMNSH
ZKZK PP [redacted] XDE
[redacted]
ZEM
S E C R E T SPOKE [redacted] SECTION FIVE [redacted] THE
FOLLOWING INFORMATION HAS BEEN PROVIDED [redacted]

[redacted] E [redacted] REACTION TO
UFO'S: [redacted]
CHASED ONE UFO.

202

T O P [redacted] U M B R A

CATEGORY = 40F M = ?

MESSAGE = 73508504

LH #0015 0167726
ZNY MMNSH
ZKZK PP [redacted] DE
[redacted]
ZEM
[redacted] SECTION 3 [redacted] THE
FOLLOWING INFORMATION HAS BEEN PROVIDED [redacted]

T O P [redacted]

203

F. RADAR TRACKING OF UFO ACTIVITIES: 1345-1402, ONE UFO (BELIEVED BALLOON) MOVED OVER AREA 160 NAUTICAL MILES 100-50 DEGREES [redacted] ALT 39,400 FT.

204

[redacted]

CATEGORY = 4Ø5 M = 2

MESSAGE = Ø328870Ø

QD #ØØ2Ø Ø7Ø15Ø1
ZNY MMNSH
ZKZK PP [redacted] DE
[redacted]

ZEM
[redacted] 2/ [redacted] 1Ø299-77 [redacted]

UNIDENTIFIED FLYING OBJECT AT 11Ø [redacted]

T O P [redacted]

205

[redacted]

CATEGORY = 405 M = 15

MESSAGE = 03007688

6. [redacted]

TO_UFO_(UNIDENTIFIED FLYING OBJECT): 0442-0512, ONE UFO (BELIEVED BALLOON) WAS TRACKED FLOATING OVER AREA 8 NAUTICAL MILES 160 DEG OF [redacted] FADED AWAY 108 NAUTICAL MILES 104 DEG OF [redacted], ALT 46, 300-50,180

T O P [redacted] E T U M B R A

206

[redacted]

= 405 M = 13

03004429

6. [redacted] RADAR TRACKING
CIFIED FLYING OBJECT: 1525-1611, ONE UFO (PROBABLY A BALLOON) MOVED
OM THE [redacted] 16 NAUTICAL MILES EAST OF [redacted] TOWARD
62,400 FT. E-2 IMPDET. WNINTEL. #0105 NNN

TOP SECRET UMBRA

207

TOP SECRET UMBRA

CATEGORY = 4Ø5 M = 11

MESSAGE = Ø3ØØ1179

6. [redacted] RADAR TRACKING OF UNIDENTIFIED FLYING OBJECTS: A. Ø652-Ø8Ø5, ONE UFO MOVED SLOWLY OVER THE AREA 22 NAUTICAL MILES EAST OF [redacted] TOWARD EAST AND FADED AFTER MOVING FOR 103 NAUTICAL MILES. B. 1115-1142, ONE UFO MOVED OVER THE AREA 11 NAUTICAL MILES EAST OF [redacted] TOWARD EAST AND FADED AFTER MOVING FOR 43 NAUTICAL MILES, ALT 43,ØØØ FT. E-2, IMPDET, WNINTEL #ØØ23
NNN

TOP SECRET UMBRA

208

M = . 9

7. [redacted] RADAR TRACKING OF <u>UNIDENTIFIED FLYING OBJECT</u>: UFO MOVED OVER THE VICINITY OF [redacted] IN [redacted] 33,000 FT. E-2, IMPDET, WNINTEL. #0067

CATEGORY = 4Ø5 M = 7

MESSAGE = Ø2995Ø62

(S.)

[redacted] RADAR TRACKING OF UNIDENTIFIED FLYING OBJECT: 1235-1258, ONE UFO MOVED SLOWLY FROM 16 NAUTICAL MILES SE OF [redacted] TOWARD EAST, ALT 26,2ØØ FT. (E2 IMPDET. WNINTEL. #ØØ57 NNN

TOP SECRET UMBRA

CATEGORY = 7 M = 3

MESSAGE = Ø2991631

[redacted]

5 [redacted] RADAR
TRACKING OF UNIDENTIFIED FLYING OBJECT: 1159-1237, ONE UFO MOVED FROM 16
NAUTICAL MILES EAST O[redacted], ALT 11,5ØØ-25,ØØØ FT. E2 IMPDET. WNINTEL. #ØØ75

TOP SECRET UMBRA

211

[redacted]

T?QIRY = 4?? ? = 10

?SSAGE = 2874?4-3

?ZCZC?169 ''''''::::

[redacted]

FM

[redacted] SECTION ONE OF FOUR [redacted] (IN PARAS SO MARKED)

SMM

[redacted]

ACC

[redacted]

YESTERDAY AFTERNOON AN UNIDENTIFIED FLYING OBJECT WAS ?ELECTED APPROXIMATELY 5? NAUTICAL MILES [redacted] MOVING IN A ?OUTHEAST[redacted] DIRECTION. THE OBJECT WAS PROBABLY A BALLOON [redacted]

T [redacted]

212

TOP SECRET UMBRA

ATEGORY = 4[illegible] M = 3

SSAGE = [illegible]21-3912[illegible]

[redacted]

XCC

[redacted]

T[illegible] SECRET UMBRA

213

UNIDENTIFIED FLYING OBJECT.11

214

~~TOP SECRET UMBRA~~

TEGORY = 4¢5 M = 5

SSAGE = 14174669

[redacted]

DIRNSA

M

[redacted]

FLYING SAUCERS REPORTEDLY OBSERVED FLYER

[redacted], SPACE OBJECTS
RE OBSERVED TRAVELING A SOUTHEASTERN TO NORTHEASTERN DIRECTION. THE POSSIBLE
AFT NUMBERED 14, IN GROUPS OF THREE WITH A FASTER LARGER OBJECT ON EACH SIDE
THE SHAPE OF ARTIFICIAL FLYING SAUCERS. [redacted] THE ABOVE
TILY DISAPPEARED OVER THE HORIZON. [redacted]

[redacted]

~~TOP SECRET UMBRA~~

215

[redacted]

ATEGORY = ALF M = 2

SSAGE = 0042-221

[redacted]

NZK FP [redacted] OFP

[redacted]

EM

S [redacted]

XKM

[redacted] UNIDENTIFIED FLYING OBJECT SIGHTED NORTHEAST OF [redacted]

NCC

[redacted] FACILITIES REFLECTED A [redacted] REPORTING AN UNIDENTIFIED FLYING OBJECT FLYING NORTH-EAST OF [redacted] TOLD [redacted] THAT HE HADSEEN AN OBJECT GOING UP AND DOWN VERTICALLY, WHICH WASIMPOSSIBLE TO BE AN AIRCRAFT. IT LOOKED LIKE A LARGE STAR. [redacted] THE OBJECT WAS GOING UP AND DOWN AT A FAST SPEED AND AT A VERY HIGH ALTITUDE, AND THATIT KEPT TURNING ROUND. [redacted] STATED HE SAWTHE OBJECT PERFECTLY AND IDENTIFIED IT AS HAVING "WHITELUMINOUS LIGHT" THAT WAS A LITTLE "BLUISH" IN COLOR. [redacted] .THERE WERE NO REFLECTIONS OF ANY OTHER [redacted] IN THIS AREA DURING THESE TIMES, WHICH MIGHT HAVE CORRELATEDTO THE UNID OBJECT THAT THE [redacted] BELIEVE THIS MAY HAVE BEEN THE SAME OBJECTNOTED FLYING TO THE SOUTH OF [redacted]

[redacted]

216

TOP SECRET UMBRA

CATEGORY = 15 M = 4

MESSAGE = Ø124Ø265

[redacted]

PP [redacted] DE

ZKZK PP [redacted] DE

[redacted]

ZEM

TOP SECRET UMBRA [redacted] NOFORN SECTION ONE OF FOUR [redacted] THE FOLLOWING INFORMATION HAS BEEN PROVIDED [redacted]

5. RADAR TRACKING OF UNIDENTIFIED FLYING OBJECTS. 6.

TOP SECRET UMBRA

217

[redacted]

TEGORY = 4Ø5 M = 15

SSAGE = ØØ6Ø66Ø1

[redacted]

EM
S E C [redacted]
XMM
[redacted] SIGHT THREE
UNIDENTIFIED FLYING OBJECTS
XCC

[redacted] AT 1915 ((2Ø15Z)) YESTERDAY, THREE
UMINOUS OBJECTS WERE SEEN IN THE WESTERN PART OF THIS CITY. THE FIRST OBJECT
AS SHAPED LIKE A HORSESHOE AND WAS WHITE IN COLOR. THE OTHER TWO WERE ROUND
ND YELLOW IN COLOR. [redacted]

[redacted]

218

TOP SECRET UMBRA

ATEGORY = 425 M = T

SSAGE = 32662519

[redacted]

EM_

S E C R E T SPOKE

XMM

[redacted] SIGHTING OF

NIDENTIFIED FLYING OBJECT

XCC

[redacted] SIGHTED AN
NIDENTIFIED FLYING OBJECT OVER [redacted] 2000 HRS [redacted]
HE UFO HAD TWO YELLOW LIGHTS, WAS FLYING AT LOW ALTITUDE, AND CHANGED ITS
EADING FROM NORTH TO WEST OVER [redacted]. NO NOISE WAS HEARD. [redacted]

NNN

TOP SECRET UMBRA

219

ATEGORY = 4Ø5 M = 22 .

ESSAGE = 2Ø639Ø13

[redacted]

EM

[redacted]

XMM

[redacted] UNIDENTIFIED FLYING OBJECT SIGHTED

XCC

1. AT 11Ø25Z [redacted] INFORMED [redacted] THAT WO LIGHTS PASSED NEAR [redacted] HEADED IN A SOUTHERLY DIRECTION, UT THAT NO NOISE WAS HEARD. SHORTLY THEREAFTER, [redacted] INFORMED EGF [redacted] THAT ACCORDING TO [redacted] THE LIGHTS WERE SEEN AGAIN TRAVELING IN A WESTERLY DIRECTION AT A DISTANCE OF APPROX 2,ØØØ FT., AND THAT NO NOISE WAS HEARD. [redacted] FROM ALL INDICATIONS THE UNIDENTIFIED OBJECT WAS IN LIGHT, ALTHOUGH NO SPECIFIC STATEMENTS TO THAT EFFECT WERE MADE. XXHH 24Ø NNNN

CATEGORY = 4.8

MESSAGE = 00683741

N = 9

[redacted]

KZK RR [redacted] DE

[redacted]

EM

[redacted]

XMM

[redacted]

LLOW## XXHH 136 NNNN [redacted] THE UFO HAD TWO

TEGORY = 4ZF M = .12

SSAGE = 30602374

[redacted]

ZK RR [redacted] DE

[redacted]

M

[redacted] SAVINT

MM

[redacted] UNIDENTIFIED FLYING OBJECTS SIGHTED

CC

N [redacted]
GHTINGS OF UNIDENTIFIED FLYING OBJECTS. THE [redacted] INFORMED
E [redacted] HAT TWO UNIDENTIFIED OBJECTS HAD BEEN SEEN AT
[redacted] OVER THE WESTERN SECTION OF [redacted] HE DESCRIBED THE FIRST AS
VING A LUMINOUS RADIATION OF 20 METERS EXTENDING IN A SPIRAL FORM FROM A
ACK CENTER. THE APPROXIMATE ALTITUDE OF THE OBJECT WAS GIVEN AS BEING BETWEEN
REE AND FOUR THOUSAND METERS, AND IT WAS STATED THAT IT LATER GAINED ALTITUDE
D WENT IN AN EASTERLY DIRECTION WITHOUT NOISE. ACCORDING TO THE [redacted], A
COND OBJECT, FOLLOWING THE SAME TRAJECTORY, WAS SIGHTED LATER AT A HIGHER
TITUDE. THE [redacted] OBSERVERS SUPPOSED THESE OBJECTS TO BE GUIDED MISSILES.
IE [redacted] REPORTED SIMILAR SIGHT- INGS BUT DESCRIBED
IE SECOND OBJECT AS A FALLING STAR. HE ALSO REPORTED THE SIGHTING OF A THIRD
JECT WHICH APPEARED TO BE A DIRIGIBLE AND WAS SEEN TRAVELING FROM NORTH TO
UTH AT A VERY HIGH ALTITUDE AND A LOW VELOCITY AT [redacted] AT [redacted]
THE [redacted] INFORM- ED THE [redacted]
THAT TWO UNIDENTIFIED FLYING OBJECTS HAD BEEN SIGHTED OVER
[redacted] HE SAID THAT THE OBJECTS WERE SEEN MOVING IN AN EASTERLY DIRECTION AT
[redacted]. HE REPORTED THAT ONE OF THE OBJECTS DESCENDED, IN A SPIRAL
SHION, TO AN ALTITUDE OF APPROXIMATELY 2500 METERS AND RADIATED AN INTENSE

[redacted]

LUE LIGHT. THESE OBJECTS WERE REPORTEDLY VISIBLE FROM ONE TO FIVE MINUTES. ONE
ALF HOUR LATER, [redacted] SAW A THIRD OBJECT TRAVELING IN THE SAME
IRECTION AND THEY PRESUMED THIS TO BE A SATELITE. (XXHH 420 NNNN)

T [redacted]

223

T [redacted]

TS'ORY = 485 M = 0

SSAGE = 0858869 0

[redacted]

ZK RR [redacted] DE

[redacted]

SIGHTS UFO

[redacted] SIGHTED A UFO DESCRIBED AS SPHERICAL OR DISC-LIKE IN FORM WITH
ESTABLISHED COLOR, BRIGHTER THAN THE SUN, WITH A DIAMETER OF ONE-HALF THE
ISIBLE SIZE OF THE MOON. 2. AT THE TIME OF OBSERVATION, OBJECT WAS ABOVE THE
PER EDGE OF THE CLOUDS ON TRUE BEARING 120 DEGREES, AZIMUTH 080 DEGREES
RAVELING NORTH. FURTHER OBSERVATION WAS IMPOSSIEL DUE TO THICK CLOUD COVER.
[redacted] ~~SECR~~ET XXHH 213 NNNN

~~T O P S E C R E T U M B R A~~

224

[redacted]

CATEGORY = 405 M = 17

MESSAGE = 03078334

[redacted]

ZNY MMNSH

ZKZK PP [redacted] DE

[redacted]

ZEM

[redacted] FINAL SECTION OF THREE [redacted]

[redacted]

225

~~TOP SECRET UMBRA~~

A. BETWEEN 2348Z [redacted] AND Ø441Z [redacted] AN UNIDENTIFIED FLYING OBJECT WAS NOTED ACTIVE OVER THE [redacted] AS FOLLOWS: 1. AN UNIDENTIFIED AIRCRAFT APPEARED 12Ø KILOMETERS [redacted] T 2348Z AND TRACKING OF THIS AIRCRAFT CEASED 4Ø KILOMETERS [redacted] AT ØØ13Z. 2. AN UNIDENTIFIED AIRCRAFT APPEARED 115 KILOMETERS EAST OF [redacted] AT Ø1Ø1Z AND TRACKING OF THIS AIRCRAFT CEASED AT 1Ø5 KILOMETERS NORHEAST OF [redacted] AT Ø1Ø5Z. 3. A [redacted] AIRCRAFT APPEARED 68 KILOMETERS EAST OF [redacted] AT HALTITUDES OF 8,ØØØ- 8,5ØØ METERS AND AT SPEEDS OF 11Ø-17Ø KNOTS. TRACKING OF THIS AIRCRAFT CEASED AT 225 KILOMETERS EAST [redacted] AT Ø441Z. IN ADDITION, AT Ø436Z, [redacted] XGDS-2

#2ØØ9 NNN

~~TOP~~ SECRET UMBRA

226

[redacted]

CATEGORY = 405 N = 7

MESSAGE = 01229969

ZCZC0CLUP87F00000000000000CAC002878 0
PP [redacted] 0CE
[redacted]
ZEM
~~QT S E C R E T UMBRA~~ [redacted] (~~ACCCTA~~ FINAL SECTION OF TWO) [redacted] ØTHE
FOLLOWING INFORMATION HAS BEEN PROVIDED BY [redacted]

~~T O P S E C R E T U M B R A~~

227

[redacted]

RADAR

TRAØCKING CF UNIDENTIFIEC FLYING CEJECTS [redacted] 1651-1655,Ø [redacted]
RADAR TØRACKEC CNE UFC MCVING ØFRCM [redacted] TC NCRTHEAST CF [redacted] ANC
FADING [redacted] 1702-1730 [redacted] RADAR TRACKEC CNE UFC MCVING ØFROM [redacted]
TC NCRTHWEST OF [redacted] ANC FACING [redacted] 1751-1755, [redacted] RADAR
TRACKED CNE UFC MCVING ØCVER SCUTHWEST C [redacted] ALTITUDE Ø11,5ØØØ
FEETØ,Ø SPEED TEN ØTC Ø2ØØ KNCTSØ.

~~TCP SECRET CFERA~~

228

[redacted]

CATEGORY = 15 M = 3

MESSAGE = Ø124Ø237

291197CZCDRA864DRUM39Ø
PP [redacted] DE
[redacted]
ZKZK PP [redacted] DE
[redacted]
ZEM
[redacted] NOFORN FINAL SECTION OF FOUR [redacted] THE
FOLLOWING INFORMATION HAS BEEN PROVIDED [redacted]

5. [redacted] RADAR TRACKING OF UNIDENTIFIED FLYING OBJECTS (UFO), A. Ø724-Ø839, [redacted] RADAR TRACKED ONE UFO MOVING FROM SOUTHEAST OF [redacted]O 21 NM NORTHWEST OF [redacted] ALTITUDE 4,6ØØ-16,ØØØ FEET. B. Ø9Ø1-Ø9Ø9, [redacted] RADAR TRACKED ONE UFO MOVING OVER 85NM 22Ø DEGREES FROM [redacted] ALTITUDE 51,ØØØ FEET. C. 1Ø11-1135, [redacted] RADAR TRACKED ONE UFO MOVING OVER [redacted] ALTITUDE 9,6ØØ FEET. D. 1323-15Ø4, [redacted] RADAR TRACKED ONE UFO MOVING [redacted] ALTITUDE 9,5ØØ FEET. E. Ø51Ø TO UNKNOWN HOURS, [redacted] RADAR TRACKEDDD ONE UFO MOVING OVER 21 NM SOUTHEAST OF [redacted]. F. [redacted] ØØ1Ø, [redacted] RADAR TRACKED ONE UFO MOVING OVER 37 NM SOUTH OF [redacted]

[redacted]

229

[redacted]

[redacted]

CATEGORY = 435 M = 1

MESSAGE = 0796873

310259CZCDR4978BCA066
PP [redacted] DE
ZNY MMCRN
ZKZK PP [redacted] DE
[redacted]
ZEM
[redacted]. THE FOLLOWING INFORMATION
HAS BEEN PROVIDED [redacted]

6. RADAR TRACKING OF UNIDENTIFIED FLYING OBJECTS.

[redacted]

230

T O P S E C R E T [redacted]

CATEGORY = 475 M = 48

MESSAGE = 73626466

LH #0006 1941546
ZNY MMNSH
ZKZK PP [redacted] DE
[redacted]
ZEM
S E C R E T SPOKE [redacted] SECTION FOUR [redacted] THE
FOLLOWING INFORMATION HAS BEEN PROVIDED [redacted]

[redacted]

231

5. RADAR TRACKING OF [UFO ACTIVITIES: 1650-1920, TWO UFO'S (BELIEVED BALLOONS) MOVED OVER NW OF [redacted] [AND COASTAL AREAS OF [redacted] [ALT 6,500-20,000 FT. A].

232

[redacted]

ATEGORY = 4Ø5 M = 2

SSAGE = ØK1Ø5373

ZK

[redacted] DE

[redacted] ,DE

[redacted] (DE)

[redacted] DE

[redacted]

M DIRNSA

[redacted]

[redacted] UNIDENTIFIED FLYING OBJECT REPORTEDLY SIGHTED

[redacted] AT 192Ø HOURS ON [redacted] AN

JECT WHICH APPEARED TO BE A ROCKET OR A (C VAL METEOR) CROSSED OVER [redacted]

OVING FROM NORTHWEST TO THE SOUTHEAST ITS TAIL LOOKED LIKE RED FIRE AND A

IERY TRAIL APPEARED IN ITS PATH FOR ABOUT THREE MINUTES NO NOISE WAS HEARD

[redacted]

[redacted]

233

T [redacted]

ATH[illegible]RY = 405   M = 15

ESSAGE = [illegible]

SGMSG012898BBC7A0729PE127MJA109TC6918
R [redacted] DE
[redacted]
O DIRNSA [redacted]
BM
S [redacted] SOURCE: [redacted] UFO SIGHTING OFF
[redacted] SAW AN ELONGATEDBALL OF FIRE MOVING
[redacted] T HIGH RATE OF SPEED. AFTER COVERING
SOME DISTANCE IT SPLIT INTO THREE BALLS OF FIRE. [redacted]

[redacted]

234

[redacted]

ATEGORY = 4,5 M = 12

SSAGE = 32244839

Z2CTC4512
ZK [redacted] DE
[redacted] DE
ADRSV ZKZK OO [redacted] DEZ
DE
[redacted]
EM
S E C R E T SAVIN
XMM
[redacted] BALLOON TRACKED
VERWATER [redacted].
XCC

BETWEEN 0415 AND 0438 ON [redacted]
[redacted] FACILITY LOCATEDAT [redacted] REFLECTED AN
UNIDENTIFIED FLYING OBJECT MOVING IN AN EASTERLY DIRECTION. THE OBJECT WAS
NITIALLY REFLECTED AT [redacted] 0415, APPROXIMATELY 50 NAUTICAL MILES
ORTHEAST OF [redacted], MOVING IN A SOUTHEASTERLY DIRECTION TO A LAST NOTED
POSITION OF [redacted] AT 0430. THE OBJECT REMAINED AT HIS LATER POSITION
TIL 0436 AND TWO MINUTES LATER THE OBJECT REPORTEDLY FADED/ON SCOPE.
ECREASING ALTITUDES WERE REFLECTED RANGING FOR 55,777TO 16,425 FEET [redacted]

T O P S E C R E T U M B R A

235

~~SECRET~~

~~TOP SECRET UMBRA~~

ATEGORY = 4Ø5 . M = 18

ESSAGE = 2627521

ZCCCJ595
KZK PP [redacted] DE
P [redacted] DF
JAMA2A3278 2292935 ZNY SSSSS RUEOJFA T DIA T DIA ARLING
ON HALL STATION RUMFSAA T USCA [redacted] T CTF 76 T [redacted] RUMORSA T
SAR SUP [redacted] RUYNUSA T CTG 73.1 P [redacted] 5524Z [redacted] FM [redacted] TO AIG NINE FOUR
EVEN BT ~~S E C R E T NOFORN~~ SECTION 2 OF 2 FORWARD WITHOUT SERVICE - INFO
ERISHABLE INTSUM NBR 33-68 (U) WEEK (SEE OUR INTSUM 32-68), THERE HAVE BEEN
ADAR SIGHTINGS FROM BOTH [redacted] FIRST RADAR SIGHTING FROM [redacted] WAS
[redacted] COMMENT: (S-NFD) STILL HAVE NO INFO TO CONFIRM OR DENY
FO'S: HOWEVER, FOUR RELIABLE AMERICAN OBSERVERS [redacted] HAVE REPORTED
IGHTINGS OF UFO'S BUT BELIEVE THEY ARE VISUAL SIGHTINGS OF SATELLITES OF SPACE
EBRIS. ONE OBSERVER SPENT OVER TWO YEARS AT SATELLITE TRACKING STATION WHERE
E AND OTHER TRACKING STATION PERSONNEL HAD VISUALLY OBSERVED SATELLITES
ASSING OVERHEAD. HE INDICATED CURRENT OBSERVATION, MADE FROM [redacted]
UTSIDE [redacted], WAS OF SAME NATURE AS HIS PREVIOUS OBSERVATIONS OF SATELLITES.

~~TOP SECRET UMBRA~~

~~SECRET~~

~~US ONLY~~

236

~~SECRET~~ CONFIDENTIAL

~~T O P S E C R E T U M B R A~~

ROORY = 44F

PAGE = ..64..8.2

CCCI393
K PP [redacted] DEI
[redacted] DR
DAEA2922 2221955 TT ZNY SSSSS RUECJFA T DIA FAV T IIIA IITTGIIIIN
IICOIIIIITIOIIIIII ODBAVMFSAA T NSCA [redacted] RUMORSA T USAR SUP [redacted]
NUBA T CTG 72.1 RUMFSAA T CTF 76 P [redacted] 6347 [redacted]
[redacted]

C. UFO'S OVER [redacted] ~~(S-NFD)~~
[redacted] REPORTS THAT UNIDENTIFIED FLYING OBJECTS
(UFO'S) HAVE BEEN SIGHTED FROM [redacted] UFO'S USUALLY APPEAR AT SAME TIME DURING
AFTERNOON HRS FOR APPROX TWOHMINUTES, FLYING IN GROUPS OF TWO OR THREE, AT
FEET. MND IS PRESENTLY FORMING A FOUR MAN MILITARY ELECTRONIC TEAM TO GO
[redacted] AND CONDUCT AN INVESTI- GATION. [redacted] ALSO SUSPECTS UFO'S MAY BE SOME
M OF [redacted] DEVICE USED BY [redacted]
MENT: ~~(S-NFD)~~ OTHER INFO RECEIVED FROM [redacted] INDICATES THAT [redacted]
GHTING OF UFO ON [redacted] A TOTAL OF THIRTY-THREE HAVE BEEN SIGHTED. UFO'S
SO DESCRIBED AS YELLOW OR GREEN. APPEAR BETWEEN 1900- PAGE THREE RUADAEA2922
S E C R E T NOFORN 2130H. AND SEEM TO DRIFT OVER THE AREA. HAVE NO INFO TO
FIRM UFO SIGHTINGS FROM U.S. SOURCES. ALSO, [redacted] RADAR HAS NOT
PORTED ANY UNUSUAL RADAR CONTACTS. POSS THAT UFO'S SOME KIND OF
PHISTICATED [redacted] DEVICE APPEARS REMOTE. UNLESS AMPLIFYING

~~T O P S E C R E T U M B R A~~

~~CONFIDENTIAL~~

~~US ONLY~~

237

CONFIDENTIAL

~~TOP SECRET UMBRA~~

VIDENCE RECEIVED, MOST PLAUSIBLE EXPLANATION OF UFO'S IS THAT THEY ARE EITHER [redacted] OR [redacted] PROPAGANDA BALLONS.

~~TOP SECRET UMBRA~~

CONFIDENTIAL

ONLY

238

~~TOP SECRET~~

TEGORY = [illegible] P = 13

SSAGE = [illegible]

7Z3A1CZCCRHB9II ::::::::::
[redacted] DE
DRZN 6 2260706
ZK OD [redacted] DE
13:7Z IYH
COMTAIWANDEFCOM/US
RCXAD INFO [redacted]
M
~~T O P S E C R E T TRINE~~ [redacted] NOFORN SECTION ONE OF FIVE [redacted] 33-68 FOR PERIOD
0800H [redacted] TO [redacted] 2800H [redacted] MEETING ATTENDED BY REPS FROM: [redacted]
27 ADIV, [redacted]

~~T O P S E C R E T UMBRA~~

~~20~~ 239

MERSHIP HIGHLIGHTS. 1. AIR: A. UFO'S OVER [redacted] THERE HAVE BEEN RADAR SIGHTINGS OF UFO'S FROM [redacted] UFO'S USUALLY SIGHTED ABOUT 1930H HOURS, IN VICINITY OF [redacted] AND NORMALLY TRAVELING INTO PREVAILING WIND. MAXIMUM ESTIMATED HEIGHT BOUT 50,000 FT AND LARGEST NUMBER AT ONE TIME HAS BEEN EIGHT. [redacted] REPORTED IRST RADAR SIGHTING OF UFO'S ON [redacted]. J-2, [redacted] IS SENDING FIVE MAN LECTRONICS TEAM TO [redacted] ON [redacted] FOR APPROX TEN DAYS TO INVESTIGATE. TEAM OMPOSED OF TWO MIL ELECTRONICS EXPERTS FROM J-2, ONE MIL ELECTRONIC EXPERT ROM ELECTRONIC WARFARE SECTION J-2, ONE CIVILIAN FROM SCIENTIFIC RESEARCH ROUP, AND ONE CIVILIAN METEROLOGIST. [redacted] (CONFIDENTIAL NOFORN) COMMENT: [redacted] LAST WEEK REPORTED VISUAL SIGHTINGS OF UFO'S FROM [redacted] BEGAN [redacted] AND HAVE CONTINUED TO PRESENT (SEE OUR INTSUM 32-68, 3-C). FO'S USUALLY DESCRIBED AS YELLOW OR GREEN, APPEAR BETWEEN 1900 AND 2130H, EEMINGLY DRIFTING OVER THE AREA. POSSIBILITY THAT UFO'S SOME KIND OF OPHISTICATED [redacted] DEVICE APPEARS REMOTE THOUGH THIS IS WHAT [redacted] USPECTS. HAVE NO INFO TO CONFIRM SIGHTINGS FROM [redacted] RADAR OR SIGINT OURCES. POSSIBLY THESE ARE REFLECTIONS OF [redacted] AND/OR [redacted] PROPAGANDA ALLOONS, DEPENDING ON WIND DIRECTION AT UFO'S ALTITUDE. HOWEVER, FOUR RELIABLE XERICAN OBSERVERS [redacted] HAVE REPORTED SIGHTING UFO'S BUT BELIEVE THEY ARE ISUAL SIGHTINGS OF SATELLITES OR SPACE DEBRIS. ONE OBSERVER SPENT OVER TWO EARS AT SATELLITE TRACKING STATION WHERE HE AND OTHER TRACKING STATION ERSONNEL HAD VISUALLY OBSERVED SATELLITES PASSING OVERHEAD. HE INDICATED THIS

240

~~T O P~~ [illegible]

SERVATION, MADE FROM [redacted] OUTSIDE [redacted] WAS OF SAT NATURE, A
ELLOWISH OBJECT ON A STEADY COURSE. TIME OF SIGHTINGS AND CLEAR ATMOSPHERE
NDITIONS FIT SATELLITE. EARLY SIGHTINGS (1930 HOURS OR SO) VIEWED FOR LONGER
RIODS 900 NNNN

~~T O P S E C R E T U M B R A~~

241

This page intentionally left blank.

Unsubstantiated and Non-source Documents

Articles Referencing Federal Government Documents Not Found in Document Repositories Related to Civil Action No. 80-1562

This page intentionally left blank.

A metallic-like disk, reported to have hovered over Santa Ana, Calif., in 1965, was the subject of

U.F.O. FILES: THE UNTOLD STORY

Though officials have long denied that they take 'flying saucers' seriously, declassified documents now reveal extensive Government concern over the phenomenon.

By Patrick Huyghe

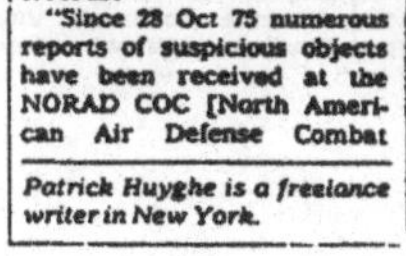

Sighting over Oregon: One of the best photographic records.

The Defense Department message bears the classification CONFIDENTIAL. "Subject: Suspicious Unknown Air Activity." Dated Nov. 11, 1975, it reads:

"Since 28 Oct 75 numerous reports of suspicious objects have been received at the NORAD COC [North American Air Defense Combat Operations Center]. Reliable military personnel at Loring AFB [Air Force Base], Maine, Wurtsmith AFB, Michigan, Malmstrom AFB, [Montana], Minot AFB, [North Dakota], and Canadian Forces Station, Falconbridge, Ontario, Canada, have visually sighted suspicious objects.

"Objects at Loring and Wurtsmith were characterized to be helicopters. Missile site personnel, security alert teams and Air Defense personnel at Malmstrom Montana reported object which sounded like a jet aircraft. FAA advised 'There were no jet aircraft in the vicinity.' Malmstrom search and height finder radars carried the object between 9,000 ft and 15,600 ft at a speed of seven knots. ... F-106s scrambled from Malmstrom could not make contact due to darkness and low altitude. Site personnel reported the objects as low as 200 ft and said that as the interceptors approached the lights went out. After the interceptors had passed the lights came on again. One hour after the F-106s returned to base, missile site personnel reported the object increased to a high speed, raised in altitude and could not be discerned from the stars....

"I have expressed my concern to SAFOI [Air Force Information Office] that we come up soonest with a proposed answer to queries from the press to prevent overreaction by the public to reports by the media that may be blown out of proportion. To date efforts by Air Guard helicopters, SAC [Strategic Air Command] helicopters and NORAD F-106s have failed to produce positive ID."

Patrick Huyghe is a freelance writer in New York.

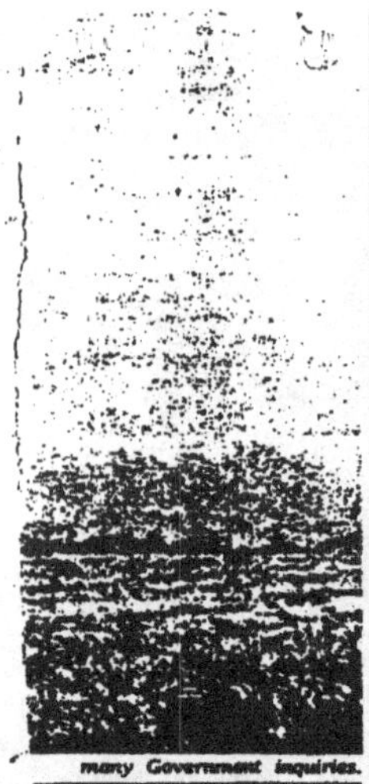

many Government inquiries.

Numerous daily updates :ept the Joint Chiefs of Staff ıformed of these incursions ›y U.F.O.'s in the fall of 1975. :epresentatives of the De- :nse Intelligence Agency and he National Security Agency - well as a handful of other ›overnment desks received opies of the National Military Command Center's reports on .e incidents. One report said ıat an unidentified object demonstrated a clear intent . the weapons storage area." hough Air Force records :ow that the C.I.A. was noti- ed several times of these :netrations over nuclear mis- .le and bomber bases, the :ency has acknowledged only ıe such notification. Subse- ıent investigations by the Air :rce into the sightings at Lor- g Air Force Base, Maine, here the remarkable series : events began, did not reveal cause for the sightings.

Despite official pronounce- ents for decades that F.O.'s were nothing more an misidentified aerial ob- :ts and as such were no .use for alarm, recently de- .ssified U.F.O. records from : C.I.A., the F.B.I. and other :deral agencies indicate at, ever since U.F.O.'s made :ir appearance in our skies the 1940's, the phenomenon ıs aroused much serious be- :nd-the-scenes concern in of- :ial circles. Details of the in- lligence community's project of U.F.O.'s have emerged over the past few years with the release of long-withheld Government records obtained through the Freedom of Information Act. Though these papers fail to resolve the U.F.O. enigma, they do manage to dispel many popular notions about the U.F.O. controversy, as well as give substance to a number of others.

Official records now available appear to put to rest doubts that the Government knew more about U.F.O.'s than it has claimed over the past 32 years. From the start, it has been convinced that most U.F.O. sightings could be explained in terms of misidentified balloons, cloud formations, airplanes, ball lightning, meteors and other natural phenomena.

But the papers also show that the Government remains perplexed about the nagging residue of unexplained U.F.O. sightings, which amount to approximately 10 percent of all U.F.O. sightings reported. Do they pose a threat to national security? Are they just a funny-looking cover for an airborne Soviet presence? Even the possibility that these *unknowns could* be evidence of extraterrestrial visitations has been given serious attention in Government circles.

While official interest in U.F.O.'s has long been thought to be strictly the concern of the Air Force, the bulk of whose records has been open to public view for nearly a decade, the recently released papers on U.F.O.'s indicate otherwise. The Departments of the Army, Navy, State and Defense, and the Defense Intelligence Agency, the National Security Agency, the Joint Chiefs of Staff, the F.B.I., the C.I.A. and even the Atomic Energy Commission produced U.F.O. records over the years. Many of these agencies still do, and many of their documents remain classified. But it is the C.I.A. that appears to have played the key role in the controversy, and may even be responsible for the Government's conduct in U.F.O. investigations throughout the years.

U.F.O.'s have been the province of the nation's intelligence community ever since the beginning of the cold war, when the notion took hold that some flying saucers might actually represent a secret, technologically advanced, foreign weapons system. "Every time we were concerned," recalls Herbert Scoville Jr., a former chief of the C.I.A.'s Office of Scientific Intelligence, "it was because we wanted to know: Did the Russians do it?"

As the cold war gave rise to the fears of the McCarthy era, [illegible] even [illegible] to the surveillance of sever[illegible] [illegible]ivate U.F.O. organizations—(as many of their members have long insisted) and to the scrutiny of dozens of individuals suspected of subversive U.F.O. activities.

Perhaps most telling of all, the Government documents on U.F.O.'s reveal that despite official denials to the contrary, Federal agencies continue to monitor the phenomenon to this day.

The monumental task of unearthing the newest batch of records on U.F.O.'s from a bureaucracy that has for years denied their existence can be traced to the efforts of a handful of inquisitive individuals who, armed with the Freedom of Information Act, set off in the mid-70's on a paper chase of U.S. Government documents on U.F.O.'s. They include Bruce S. Maccabee, a Silver Spring, Md., physicist working for the Navy, who has managed to obtain the release of more than 1,300 pages of documents on U.F.O.'s from the F.B.I.; W. Todd Zechel of Prairie du Sac, Wis.; Robert Todd of Ardmore, Pa.; Larry W. Bryant of Arlington, Va.; and Brad C. Sparks, a student in astrophysics at Berkeley whose five-year pursuit of the C.I.A.'s U.F.O. file eventually provided the foundation for a ground-breaking Freedom of Information lawsuit filed by Ground Saucer Watch (G.S.W.), an Arizona-based U.F.O. organization.

At the request of G.S.W. director William H. Spaulding, Peter Gersten, an attorney in the New York firm of Rothblatt, Rothblatt & Seijas, filed a civil action against the C.I.A. in December 1977 demanding all U.F.O. records in the agency's possession. The suit seemed to have achieved its goal when late last year the agency released about 400 documents — nearly 900 pages of memos, reports and correspondence that attest to the agency's long involvement in U.F.O. matters. But the civil action has not seen its final day in court.

By Gersten's account, the agency has arbitrarily withheld documents, made deletions without merit, and failed to conduct a proper search for U.F.O. materials. The agency's current actions, he says, perpetuate its 30-year policy of deliberate deception and dishonesty about U.F.O.'s. "What has been released to us seems to have been rather carefully selected," says Gersten. "We suspect that the agency is withholding at least 200 more documents than the 57 they have admitted they are keeping from us to protect intelligence sources." Victor Marchetti, a former executive deputy director, agrees with Gersten. "[illegible] entire exercise," Marchetti wrote recently in a magazine article, "has the same aroma of the agency's previous messy efforts to hide its involvement in drugs and mind-control operations, both prime examples of a successful intelligence cover-up."

□

The first sighting to be labeled a "flying saucer" by the press occurred on June 24, 1947, when an Idaho businessman flying his plane near Mount Rainier observed nine disc-shaped objects making undulating motions "like a saucer skipping over water." As early as World War II, Allied bomber pilots had told of "balls of light" that followed their flights over Japan and Germany. A U.S. Eighth Army investigation concluded that they were the product of "mass hallucination."

These and other incidents were reported in a 1973 book by David Michael Jacobs, "The UFO Controversy in America," which until the recent release of Government documents was the most comprehensive reconstruction of the Government's U.F.O. involvement.

When Scandinavians reported cigar-shaped objects in 1946, U.S. Army intelligence suspected that the Russians had developed a secret weapon with the help of German scientists from Peenemünde. The C.I.A., then known as the Central Intelligence Group, secretly began keeping tabs on the subject.

When the unknown objects returned to the skies, this time over the United States in the summer of 1947, the Army Air Force set out to determine what the objects were. Within weeks, Brig. Gen. George F. Schulgen of Army Air Corps Intelligence requested the F.B.I.'s assistance "in locating and questioning the individuals who first sighted the so-called flying discs. . . ." Undoubtedly swayed by flaring cold-war tensions, Schulgen feared that "the first reported sightings might have been by individuals of Communist sympathies with the view to causing hysteria and fear of a secret Russian weapon." J. Edgar Hoover agreed to cooperate but insisted that the bureau have "full access to discs recovered."

The Air Force's behind-the-scenes interest contrasted sharply with its public stance that the objects were products of misidentifications and an imaginative populace. A security lid was imposed on the subject in July 1947, hiding a potentially "embarrassing situation" the following month, when both the Air Force and the F.B.I. began suspecting they might actually be secret weapons. High-level reassurances were ob[illegible]ed that this was not so.

By the end of the summer, the F.B.I. had "failed to reveal any indication of subversive individuals being involved in any of the reported sightings." A RESTRICTED Army letter that found its way to Hoover's desk said that the bureau's services actually had been enlisted to relieve the Air Forces "of the task of tracking down all the many instances which turned out to be ashcan covers, toilet seats and whatnot." Incensed, Hoover moved quickly to discontinue the bureau's U.F.O. investigations.

In September of that year, the Commanding General of the Army Air Force received a letter from the Army Chief of Staff Lieut. Gen. Nathan F. Twining, saying that "the phenomenon reported is of something real and not visionary or fictitious," that the objects appeared to be disc-shaped, "as large as man-made aircraft," and "controlled either manually, automatically or remotely." At Twining's request, project "Sign" was established.

"Sign" failed to find any evidence that the objects were Soviet secret weapons and before long submitted an unofficial "Estimate of the Situation," classified TOP SECRET, which indicated that U.F.O.'s were of interplanetary origin. The estimate eventually reached Air Force Chief of Staff Gen. Hoyt S. Vandenberg, who rejected it for lack of proof. "Sign's" inconclusive final report remained classified for the next 12 years.

After "Sign," the Air Force continued to collect U.F.O. data under the code name "Grudge." This six-month project found no evidence of foreign scientific development and therefore no direct threat to national security. It did, however, stress that the reported sightings could be dangerous. "There are indications that the planned release of related psychological propaganda would cause a form of mass hysteria," the report stated. "Employment of these methods by or against an enemy would yield similar results . . . governmental agencies interested in psychological warfare should be informed of the results of this study."

A press release following the termination of "Grudge" allowed the public to believe that the Air Force was no longer interested in U.F.O.'s. But the Air Force continued to collect reports through normal intelligence channels until a dramatic sighting of a U.F.O. at the Army Signal Corps radar center in Fort Monmouth, N.J., in 1951 led to the reactivation of "Grudge," [illegible]e Air Force project was renamed "Blue Bo[illegible] in 1952, a year that saw a record number of U.F.O. reports.

□

The situation got out of hand during the summer of 1952. On the morning of July 28, the Washington Post revealed that U.F.O.'s had been tracked on radar at Washington National Airport, the second such incident in a week. Reporters stormed Air Force headquarters in the Pentagon, where switchboards were jammed for days with U.F.O. inquiries. Military installations across the country handled such a volume of reports that "regular intelligence work had been affected," reported The New York Times.

These events prompted action at C.I.A. headquarters, apparently at a request "from the Hill." From the start, the agency's involvement was to be kept secret. An August 1 C.I.A. memo recommended that "no indication of C.I.A. interest or concern reach the press or public, in view of their probable alarmist tendencies to accept such interest as 'confirmatory' of the soundness of 'unpublished facts' in the hands of the U.S. Government."

The C.I.A.'s Office of Scientific Intelligence (O.S.I.) found that the Air Force's investigation of the U.F.O. phenomenon was not sufficiently rigorous to determine the exact nature of the objects in the sky. Neither did the Air Force deal adequately with the potential danger of U.F.O.-induced mass hysteria, or the fact that our air vulnerability was being seriously affected by the U.F.O. problem. O.S.I. chief H. Marshall Chadwell thought that our nation's defenses were running the increasing risk of false alert and, worse yet, "of falsely identifying the real as phantom." He suggested that a national policy be established "as to what should be told the public" and, furthermore, that immediate steps be taken to improve our current visual and electronic identification techniques so that "instant positive identification of enemy planes or missiles can be made." Ever vigilant, the C.I.A. was keeping an eye on the possibility that U.F.O.'s could be of Soviet origin.

By the winter of 1952, Chadwell had drafted a National Security Council proposal calling on a program to solve the problem of instant positive identification of U.F.O.'s. In a memo that accompanied the proposal, Chadwell urged that the reports be given "immediate attention." He thought that "sightings of unexplained objects at great altitudes and traveling at high speeds in the vicinity of major U.S. defense installations are of such nature that they are not attributable to natural phenomena or known types of aerial vehicles." He said that O.S.I. was proceeding with the establishment of a consulting group "of sufficient competence and stature to . . . convince the responsible authorities in the community that immediate research and development on this subject must be undertaken."

But C.I.A. Director Gen. Walter B. Smith's interest apparently lay elsewhere. In a letter to the Director of the Psychological Strategy Board, he expressed a desire to discuss "the possible offensive and defensive utilization of these phenomena for psychological warfare purposes." Only later did Director Smith authorize recruiting an advisory committee of outside consultants.

□

The scientific panel met for four days beginning Jan. 14, 1953. Chaired by Dr. H.P. Robertson, an expert in physics and weapons systems, the panel essentially bestowed the scientific seal of approval on previously established official policy regarding U.F.O.'s. The distinguished panelists felt that all the sightings could be identified once all the data were available for a proper evaluation — in other words, the phenomena—according to the panel's r[illegible] were not "beyond the dom[illegible] of present knowledge of physical sciences." Neither did the panelists find U.F.O.'s to be a direct threat to national security, though they believed that the volume of U.F.O. reports could clog military intelligence channels, precipitate panic, and lead defense personnel to ignore real indications of hostile action. The panel worried about Soviet manipulation of the phenomenon; that the reports could make the public vulnerable to "possible enemy psychological warfare." The real danger, they concluded, was the reports themselves.

Fearing that the myth of U.F.O.'s might lead to inappropriate actions by the American public, the panelists decided that a "broad educational program integrating efforts of all concerned agencies" must be undertaken. They sought to strip U.F.O.'s of their "aura of mystery" through this program of "training and 'debunking.'" The program would result in the "proper recognition of unusually illuminated objects" and in a "reduction in public interest in 'flying saucers.'" The panelists recommended that their mass-media program have as its advisers psychologists familiar with mass psychology and advertising experts, while Walt Disney Inc. animated cartoons and such personalities as Arthur Godfrey would help in the educational drive. To insure complete control over the situation, the panel members suggested that flying-saucer groups be "watched because of their potentially great influence on mass thinking if widespread sightings should occur. The apparent irresponsibility and the possible use of such groups for subversive purposes should be kept in mind."

The panel's recommendations called for nothing less than the domestic manipulation of public attitudes. Whether these proposals were acted upon, the C.I.A. will not say. But the report was circulated among the top brass at the Air Technical Intelligence Center, the C.I.A.'s Board of National Estimates (of which Hoover was a member), the C.I.A.'s bureau chiefs, the Secretary of Defense, the chairman of the National Security Resources Board, and the director of the Federal Civil Defense Administration, who eventually sent a representative to meet with C.I.A. officials in order to "implement the appropriate aspects of the Panel's Report as applicable to Civil Defense."

The Government's efforts in the 50's and 60's to squelch

U.F.O.'s went beyond debunking and ... touched the fiber of constitutionally protected free speech. According to author David Michael Jacobs, in 1953 the Air Force pressured Look magazine into publishing disclaimers throughout an article by retired Maj. Donald E. Keyhoe entitled "Flying Saucers From Outer Space." Then again, in 1965, the Army — in a prepublication review — denied clearance for a U.F.O.-related article by one of its employees, Larry W. Bryant, a technical editor, until he took the issue to court.

Meanwhile, the C.I.A. and the F.B.I. proceeded routinely in the surveillance of U.F.O. organizations and U.F.O. enthusiasts. People with U.F.O. interests were checked out by the F.B.I. at the request of the C.I.A., the Air Force, or private citizens inquiring about possible subversive activities. None caused as much consternation as the case of Major Keyhoe and the organization he directed, the National Investigations Committee on Aerial Phenomena (NICAP).

The C.I.A. appears to have had a protracted interest in NICAP, which was founded in 1956 and utilized by Keyhoe as an organizational tool for challenging the alleged Air Force cover-up on U.F.O.'s. Both the C.I.A. and the Air Force were upset by NICAP's wide-ranging influence. Its prestigious board of directors included, among others, Vice Adm. Roscoe Hillenkoetter, the first C.I.A. Director (1947-1950). "The Air Force representatives believe that much of the trouble . . . with Major Keyhoe . . . could be "alleviated," states a C.I.A. memo dated May 16, 1958, "if the Major did not have such important personages as Vice Admiral R. H. Hillenkoetter, U.S.N. (Ret.) . . . on the board. . . ." The Air Force suggested that if the Admiral were shown the SECRET panel report he might understand and take "appropriate actions." Whether or not the Air Force got through to the admiral, Hillenkoetter resigned from NICAP in 1961.

The 60's saw further C.I.A. interest in NICAP. After a flurry of Washington-area sightings in 1965, the agency contacted NICAP about seeing some of its case files on the matter. Richard H. Hall, then NICAP assistant director, chatted with a C.I.A. agent in the NICAP office about the sightings, NICAP's methodology, and Hall's background. The agent's memo on the visit suggests that the C.I.A. had some role in mind for Hall, predicated upon his being granted a security clearance. Nothing apparently came of the suggestion. A later set of

The New York Times

C.I.A. paper reveals an interest in NICA organizational structure and notes that "this group included some ex-C.I.A. and Defense Intelligence types who advise on investigative techniques and NICAP-Government relations." There are presently three former C.I.A. employees on the NICAP board of directors, including Charles Lombard, a congressional aide to Senator Barry Goldwater, who is himself a NICAP board member; and retired U.S. Air Force Col. Joseph Bryan III. Bryan feels, as he did back in 1959 when he joined the board, that U.F.O.'s are interplanetary. NICAP's current president is Alan Hall, a former C.I.A. covert employee for 30 years.

□

In 1966, mounting discontent from members of the press, Congress and the scientific community compelled the Air Force to commission an 18-month scientific study of U.F.O.'s under the direction of Edward U. Condon, professor of physics at the University of Colorado. The politically expedient study, in which one-third of the 91 cases examined remained unidentified, reiterated official policy with one novel twist: U.F.O.'s "educationally harmed" schoolchildren who were allowed to use science study time to read books and magazine articles about U.F.O.'s. Condon wanted teachers to withhold credit from any student U.F.O. project. The Air Force took the cue and disbanded project "Blue Book" in 1969.

Less than a decade later, the White House, perhaps in an attempt to make good Jimmy Carter's campaign promise to tell all about U.F.O.'s, suggested via science advisor Frank Press that possibly NASA could undertake a review of any significant new findings since Condon's study. NASA examined the offer, but saw no way to attack the problem on a scientific basis without physical evidence. They envisioned a public-relations nightmare if they were to accept such a project, and so rejected it. A frank, in-house evaluation of NASA's options, however, noted that a hands-off attitude only begged the question. So in good spirit, the space agency offered to examine any piece of physical evidence brought to its attention. That position led one Federal aviation official to comment: "If you get a piece of the thing, fine. But don't bother me with anything else."

□

These days, the Air Force admits to nothing more than a "transitory interest" in the phenomenon, although military directives still exist for reporting U.F.O.'s.

The C.I.A. is still wary of the possibility that U.F.O.'s may be of Soviet origin. "The agency's interest," says Katherine Pherson, a public-affairs officer for the C.I.A., "lies in its responsibility to forewarn principally of the possibility that a foreign power might develop a new weapons system that might exhibit phenomena that some might categorize as a U.F.O. But there is no program to actively collect information on U.F.O.'s." The agency's interest cannot be denied, however, as two 1976 memos reveal.

The first, dated April 26, states: "It does not seem that the Government has any formal program in progress for the identification/solution of the U.F.O. phenomena. Dr. [name deleted] feels that the efforts of independent researchers, [phrase deleted], are vital for further progress in this area. At the present time, there are offices and personnel within the agency who are monitoring the U.F.O. phenomena, but again, this is not currently on an official basis."

Another memo, dated July 14, and routed to the deputy chief in the Office of Development and Engineering, reads: "As you may recall, I mentioned my own interest in the subject as well as the fact that DCD [Domestic Collection Division] has been receiving U.F.O. related material from many of our S & T [Science and Technology] sources who are presently conducting related research. These scientists include some who have been associated with the Agency for years and whose credentials remove them from the 'nut' variety."

□

If nothing else, the success of the U.F.O. paper chase may have lent U.F.O.'s a measure of respectability that has eluded the subject for the past third of a century. Though it appears that no U.F.O. sighting has ever represented an airborne Soviet or foreign threat, the possibility that such an event could occur remains foremost in the cold-war-conscious Government mind. Should that threat come to pass, military officials believe, our nation's sophisticated defense system would know about it before someone getting a glass of milk in the middle of the night sees the threat hovering outside the kitchen window. Or so we are made to understand the Air Force's seemingly nonchalant advice to the public: "If you see a U.F.O. and you feel the situation warrants it, call your local police." ■

Govt.'s Super-Secret Security Agency Warns:

Take UFOs Seriously or Be Prepared for Sneak Invasion By Space Aliens

27Nov79
"National Enquirer"

In one of the most startling government reports on UFOs ever, America's super-secret National Security Agency (NSA) says UFOs are real — and warns that the country should prepare for a confrontation with space aliens.

The sensational document — obtained exclusively by The ENQUIRER — rips apart the notion that all UFOs are hoaxes or hallucinations.

And what's more, the NSA has sounded a bone-chilling alarm:

If America doesn't start taking the sightings seriously, we are leaving ourselves wide open to the possibility of a Pearl Harbor-type UFO invasion!

"The very fact that UFO phenomena have been witnessed all over the world from ancient times, and by considerable numbers of reputable scientists in recent times, now indicates rather strongly that UFOs are not all hoaxes," declares the dramatic report.

"And if anything, rather than diminishing, the modern trend is toward increased reports, from all sources."

In one three-month period alone, "Air Force records show 35 sightings whose nature could not be determined," reveals the study.

The fascinating document was prepared by the NSA in 1968.

Initial requests to have the report released were flatly rejected and it was obtained by The ENQUIRER only through appeals under the Freedom of Information Act. Even then, we were given a heavily censored version of the startling study.

But the parts of the report released leave little doubt that as early as 1968 the nation's most secret intelligence agency was deeply concerned about UFOs.

Smashing to bits the idea that all UFOs are hallucinations, the report clearly states:

• "A considerable number of instances exist in which there are groups of people and a radar or radars seeing

DANGEROUS THREAT: The high percentage of sightings by reputable people and tho the two spacecraft hovering over Dulce, N. Mex. (photo at left), the trio of flying saucers and the dazzling four-pointed UFO over Johannesburg, South Africa (photo at right), that UFOs are not all hoaxes" . . . and they represent a real threat to our nation's su

DANGEROUS THREAT: The high percentage of sightings by reputable people and those captured on film, like the two spacecraft hovering over Dulce, N. Mex. (photo at left), the trio of flying saucers over Italy (center photo), and the dazzling four-pointed UFO over Johannesburg, South Africa (photo at right), "indicates rather strongly that UFOs are not all hoaxes" . . . and they represent a real threat to our nation's survival.

the same thing at the same time.

- "On occasion, physical evidence of a circumstantial nature was reported to have been found to support witnessed sightings.

- "A continuing high percentage of reports of unusual aerial objects are being reported by people in responsible positions in science, government and industry."

The report notes, in no uncertain terms, that some eminent scientists believe that UFOs are extraterrestrial in origin — and cautions that this fact "cannot be disregarded."

Moreover, there could be some very "serious" and "far-reaching human survival implications," warns the report, adding:

"If 'they' discover you, it is an old but hardly invalid rule of thumb, 'they' are your technological superiors.

"Human history has shown us time and again the tragic results of a confrontation between a technologically superior civilization and a technologically inferior people."

In an attempt to rally concern about the presence of UFOs, the report pleads for stepped-up action.

"Up until this time, the leisurely scientific approach has too often taken precedence in dealing with UFO questions," it states.

Then, making a frightening comparison to highlight the awesome dangers, the report reasons:

"If you are walking along a forest path and someone yells, 'rattler,' your reaction would be immediate and defensive. You would not take time to speculate before you act. You would have to treat the alarm as if it were a real and immediate threat to your survival.

"Investigation would become an intensive emergency action to isolate the threat and to determine its precise nature. It would be geared to developing adequate defensive measures in a minimum amount of time.

"It would seem a little more of this survival attitude is called for in dealing with the UFO problem."

The report — which was drastically censored by the NSA to protect national security — doesn't indicate who wrote it or for whose eyes it was intended.

Although it states that the nature and origin of UFOs are still a mystery, it sounds a clear alert. The potential threat is dramatically underscored in the document — by inclusion of a chilling list of other instances in history when nations failed to prepare adequately in the face of danger.

And the list of instances, called "Other Examples of Blindness to Surprise Material Causing Defeat," starkly points out:

"Because we could not appreciate the power of the carrier strike force, we were surprised and defeated at Pearl Harbor.

"Because Americans found it incomprehensible that a crude technology could effectively defend itself against a sophisticated weapons system, many aircraft were lost to World War 2 model antiaircraft and small arms fire in Vietnam."

The implication is unmistakable: Only if we are fully prepared to meet the potential threat from UFOs can we ever rest safe and secure as a nation.

— *THOMAS L. MULDOON*

NATIONAL ENQUIRER (USPS 372-940) Published weekly by National Enquirer, Inc., Lantana, Fla. [illegible], Vol. [illegible], No. [illegible]. MAIL SUBSCRIPTIONS [illegible] a year in U.S., [illegible] a year in Canada, [illegible] a year for foreign addresses. Second Class postage paid at Lake Worth, Fla., and at additional mailing offices.

cont. from prev. page

76 Societe Vauclusienne d'Etude Des Phenomenes Spatiaux
El Pontet
77 Union des Groupements Esplologiques de France et des Pays de Langue Francaise
Valence
78 Verification et Etude des Ovni pour Nimes et la Contree Avoisante (VERONICA)
Nimes

SPAIN

79 Centro de Estudios Interplanetarios (CEI)
Barcelona
80 Circulo de Estudios Sobre Objectos no Identificados
Valencia

PORTUGAL

81 Centro de Estudos Astronomicos e de Fenomenos Insolitos
Lisbon

BELGIUM

82 Groupement pour l'Etude des Sciences d'avant-garde
Brussels
83 Societe Belge d'Etude des Phenomenes Spatiaux (SOBEPS)
Brussels

NETHERLANDS

84 Nederlands UFO Onderzoek Bureau
The Hague
85 NOBOVO
Uithuizermeeder

NORTHERN IRELAND

86 UFO Research Centre
Armagh

ITALY

87 Centro UFOlogico Nazionale
Milan
88 Gruppo Clypeus
Turin
89 Independent National Commission for the Study of Anomalous Aerial Phenomena (CNIFAA)
Bologna
90 La Contact International
Rome
91 S.H.A.D.O.
Genoa

DENMARK

92 Dansk UFO Center
Thisted
93 Skandinavisk UFO Information (SUFOI)
Kastrup
94 UFO Studiekreds
Copenhagen

FINLAND

95 The UFO Researchers of Finland
Turku

SCOTLAND

96 Edinburgh University UFO Research Society
Edinburgh

SWEDEN

97 Arbetsgruppen for UFOlogi
Sodertalje
98 Gotesborgs Informations Center for Oidentifierade Flygande Foremal
Goteborg
99 Swedish UFO Research Center
Nassjo
100 UFO Sverige
Motala

WEST GERMANY

101 Centrales Enforschungsnetz Aubergewohnlicher Phanomene (CENAP)
Mannheim
102 UFO-SIG
Berlin

YUGOSLAVIA

103 ODISEJA
Slovenija

JAPAN

104 CAB International
Yokohama

NEW ZEALAND

105 Civilian Saucer Investigation
Auckland
106 New Zealand Scientific Space Research Group
Auckland

AUSTRALIA

107 Australian Co-ordination Section—Centre for UFO Studies (ACOS)
Gosford, New South Wales
108 Australian Flying Saucer Research Society
Adelaide, South Australia
109 Queensland UFO Research Bureau
Brisbane, Queensland
110 Tasmanian UFO Investigation Center (TUFOIC)
Hobart, Tasmania
111 UFO Information Center (UFOIC)
Lane Cove, Victoria
112 UFO Research Network
Sydney, New South Wales
113 Victoria UFO Research Society
Moorabin, Victoria

I U Report... JAN 78

INTERNATIONAL INCIDENTS OF INTEREST

THE U.S. GOVERNMENT & THE IRAN CASE

MORE INFORMATION ON THAT IRAN CASE . . . STRAIGHT FROM THE "TOP BRASS"!

In IUR Vol. 1, No. 1, a case in the Foreign Forum feature described an encounter between Iranian Air Force jets and a UFO which played "cat and mouse" with them, appearing on radar and "paralyzing" their weapons and electronics systems when the jets attempted to open fire on it. In the interim, rumors began to circulate that a government document detailing the incident was being examined by high government officials. UFO researcher Charles Huffer took an interest in securing a copy of this document while in Germany, but all his efforts met with official rebuffs. This past summer on a trip to the United States, Mr. Huffer managed to obtain a copy of the original teletype message reproduced below which gives details of the case from the Pentagon, under the auspices of the Freedom of Information Act.

What is particularly interesting about the document is the list of official agencies which received copies of the message. Briefly translating the official acronyms, the document was received by the following offices and agencies: the Secretary of State, the Central Intelligence Agency, the White House, the Air Force and Army Chiefs of Staff, the Chief of Naval Operations, the Defense Intelligence Agency, the Commander in Chief of U.S. Naval Forces in the Middle East, the Commander in Chief of the U.S. Air Force in Europe, the European Defense Air Command, and the Commander in Chief of Forces in Europe. An impressive list!

IUR and others have been intrigued by the fact that this particular UFO report seems to have become the concern of the "top brass" in the U.S. military defense structure. Do all UFO reports undergo this sort of distribution and what happens to them once they reach these prestigious offices?

IUR contacted the local state senator's office Military Liaison Officer who was able to be of some assistance in the matter. She called the Army, Navy, and Air Force offices in Washington as well as the Department of Defense and addressed these very questions on behalf of IUR. The answers were illuminating: all these agencies agreed that the "top brass" mentioned in the distribution list were routinely informed of *every* item of interest which comes out of the sensitive Middle East area. Hence, it is not unusual at all that copies of the report should have made their way to these offices. All correspondence that makes the military wires from this area is so relayed. As to whether or not all UFO reports from *every* area are conveyed to these sources, that question must remain unanswered at this time except by implication; any report from a "hot" area would be relayed. What happens to them once they reach the top brass is another thing. IUR was not able to glean any information on whether or not there exist any files, or a normal procedure in dealing with UFO sightings.

IUR was also able to find some information on the allowed international lighting configurations and colors. The initial, primary object, according to the report had alternating blue, green, red and orange lights. FAA Federal Aviation Regulation 43 (91.1) on General Operation and Flight Applicability states that blue lights are not authorized on U.S. planes. Furthermore, the International Civil Aviation Organization states that blue lights are illegal and that only certain stringently specified light configuations are allowable on aircraft. However, IUR must point out that even in past sightings personally investigated, the witnesses have perceived the green navigational lights of planes as blue or blue-green.

VERBATIM WIRE MESSAGE

This report forwards information concerning the sighting of an UFO in Iran on 19 September 1976.

A. At about 12:30 AM on 19 Sep. 76 the ______________ received four telephone calls from citizens living in the Shemiran area of Tehran saying that they had seen strange objects in the sky. Some reported a kind of bird-like object while others reported a helicopter with a light on. There were no helicopters airborne at that time.

After he told the citizen it was only stars and had talked to Mehrabad Tower he decided to look for himself. He noticed an object in the sky similar to a star bigger and brighter. He decided to scramble an F-4 from Shahrokhi AFB to investigate.

B. At 0130 hrs on the 19th the F-4 took off and proceeded to a point about 40 NM (nautical miles—Ed.) North of Tehran. Due to its brilliance the object was easily visible from 70 miles away. As the F-4 approached a range of 25 NM he lost all instrumentation and communications (UHF and intercom). He broke off the intercept and headed back to Shahrokhi. When the F-4 turned away from the object and apparently was no longer a threat to it the aircraft regained all instrumentation and communications. At 0140 hrs a second F-4 was launched. The backseater acquired a radar lock on at 27 NM, 12 o'clock high position with the VD (rate of closure) at 150 NMPH. As the range decreased to 25 NM the object moved away at a speed that was visible on the radar scope and stayed at 25 NM.

C. The size of the radar return was comparable to that of a 707 tanker. The visual size of the object was difficult to discern because of its intense brilliance. The light that it gave off was that of flashing strobe lights arranged in a rectangular pattern and alternating blue, green, red and orange in color. The sequence of the lights was so fast that all the colors could be seen at once. The object and the pursuing F-4 continued on a course to the south of Tehran when another brightly lighted object, estimated to be one half to one third the apparent size of the moon, came out of the original object. This second object headed straight toward the F-4 at a very fast rate of speed. The pilot attempted to fire an AIM-9 missile at the object but at that instant his weapons control panel went off and he lost all communications (UHF and Interphone). At this point the pilot initiated a turn and negative G dive to get away. As he turned the object fell in trail at what appeared to be about 3-4 NM as he continued in his turn away from the primary object the second object went to the inside of his turn then returned to the primary object for a perfect rejoin.

D. Shortly after the second object joined up with the primary object another object appeared to come out of the other side of the primary object going straight down at a great rate of speed. The F-4 crew had regained communications and the weapons control panel and watched the object approach the ground anticipating a large explosion. This object appeared to come to rest gently on the earth and cast a very bright light over an area of about 2-3 kilometers. The crew descended from their altitude of 25M to 15M and continued to observe and mark the object's position. They had some difficulty in adjusting their night visibility for landing so after orbiting Mehrabad a few times they went out for a straight in landing. There was a lot of interference on the UHF and each time they passed through a mag. bearing of 150 degree from Fhrarad they lost their communications. (UHF and Interphone) and the ins fluctuated (instruments —Ed.) from 30 degrees-50 degrees. The one civil airliner that was approaching Mehrabad during this same time experienced communications failure in the same vicinity (KILO ZULU) but did not report seeing anything. While the F-4 was on a long final approach the crew noticed another cylinder shaped object about the size of a T-bird at 10M [10,000 ft.—Ed.] with bright steady lights on each end and a flasher in the middle. When queried the tower stated there was no other known traffic in the area. During the time that the object passed over the F-4 the tower did not have a visual on it but picked it up after the pilot told them to look between the mountains and the refinery.

E. During daylight the F-4 crew was taken out to the area in a helicopter where the object apparently had landed. Nothing was noticed at the spot they thought the object landed (a dry lake bed) but as they circled off the the west of the area they picked up a very noticeable beeper signal. At the point where the return was the loudest was a small house with a garden. They landed and asked the people within if they had noticed anything strange last night. The people talked about a loud noise and a very bright light like lightning. The aircraft and area where the object is believed to have landed are being checked for possible radiation.

More information will be forwarded when it becomes available.

FIRST PHOTOS OF GUATEMALA VIDEOTAPE

The UFO about to pass behind a nearby tree . . .

. . . having just emerged from behind the tree . . .

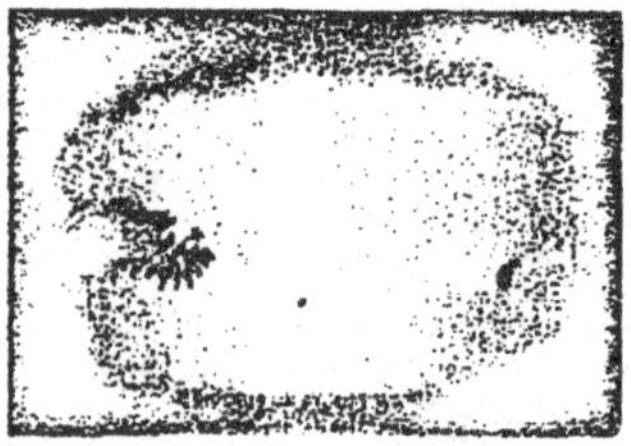

. . . and moving off into the distance.

In IUR, Vol. 2, No. 11, the Foreign Forum feature mentioned a case wherein a Guatemalan camera crew videotaping a car commercial turned their camera towards a "UFO" which fortuitously flew into their field of view. The incident took place on Oct. 5, 1977 at 10:40 AM, in Guatemala City and is the first incident, to our knowledge, of a color videotape recording a "UFO." Through the (cont. on next page)

THE NEW YORK TIMES, WEDNESDAY, AUGUST 26, 1987

Report of U.F.O. Crash in '47 Called False by Science Panel

WASHINGTON, Aug. 25 (Reuters) — Documents purported to be from the Truman White House that say the Pentagon recovered a crashed flying saucer and the bodies of four alien creatures in 1947 are "clumsy counterfeits," according to a report by a group of scientists.

The report was released Monday by the group, the Committee for the Scientific Investigation of Claims of the Paranormal. The report was prepared by Phillip J. Klass, the Washington editor of Aviation Week & Space Technology magazine and a leading debunker of reports on unidentified flying objects.

The chairman of the committee, Paul Kurtz, a University of Buffalo philosophy professor, said the documents represented "one of the most deliberate acts of deception ever perpetrated against the news media and the public."

The documents, which said President Truman created a secret unit called Majestic 12, or MJ-12, to study the saucer and its contents, were made public in May by William L. Moore, a researcher on U.F.O.'s.

White House Report

Mr. Moore told reporters then that his research team had found a key White House report in the National Archives dated July 14, 1954. It appeared to have been prepared for the Air Force by Robert Culter, a White House aide, and mentioned a change in plans for an MJ-12 briefing for President Eisenhower.

Mr. Klass said his research showed that the document was false. He said that Mr. Cutler was not in Washington when the report was supposedly written, having left for Europe 11 days earlier.

According to a National Archives memo released by Mr. Klass, the Archives is also suspicious because Mr. Moore's does not bear the required top secret registration number and is marked "Top Secret Restricted Information" — a designation that was not used until the Nixon Administration.

Another document in which Truman supposedly ordered Defense Secretary James Forrestal to create MJ-12, is also a forgery, Mr. Klass said.

The document did not follow the format Truman used in writing letters to his Cabinet secretaries and was created by superimposing a spurious message on a photograph of an authentic Truman letter, the editor concluded.

UFO GROUP WANTS SECRET DOCUMENTS
BY ROBERT SANGEORGE

WASHINGTON (UPI) -- A GROUP OF UFO BUFFS WANTS THE SUPREME COURT TO ORDER THE RELEASE OF MYSTERIOUS MATERIAL COLLECTED ABOUT PURPORTED VOYAGERS FROM OUTER SPACE AND HELD BY THE SUPER-SECRET NATIONAL SECURITY AGENCY.

A NEW YORK CITY-BASED GROUP CALLED CITIZENS AGAINST UFO (UNIDENTIFIED FLYING OBJECT) SECRECY WANTS THE JUSTICES TO ORDER THE AGENCY TO RELEASE THE 135 DOCUMENTS UNDER THE FREEDOM OF INFORMATION ACT.

THE CASE BEGAN IN LATE 1978 WHEN THE UFO GROUP FILED A FORMAL REQUEST FOR 18 AGENCY DOCUMENTS UNDER THE INFORMATION LAW. THE NATIONAL SECURITY AGENCY FLATLY REFUSED TO RELEASE THE INFORMATION, CLAIMING IT IS EXEMPT FROM DISCLOSURE UNDER THE LAW.

A FEW MONTHS LATER, THE GROUP EXPANDED ITS REQUEST TO INCLUDE "ALL DOCUMENTS IN THE POSSESSION OR UNDER THE CONTROL OF THE NATIONAL SECURITY AGENCY RELATING TO OR PERTAINING TO UNIDENTIFIED FLYING OBJECTS AND THE UFO PHENOMENA."

IN RESPONSE, THE AGENCY ADMITTED IT HAS 135 SUCH "UFO-RELATED" DOCUMENTS, BUT AGAIN DECLINED TO RELEASE THEM. IT CLAIMED SUCH DISCLOSURE WOULD REVEAL SOME OF ITS TOP SECRET ELECTRONIC MONITORING AND INTERCEPTION TECHNIQUES.

THE AGENCY IS A DEFENSE DEPARTMENT UNIT HEADQUARTERED AT FORT MEADE, MD. ONE OF ITS MAIN RESPONSIBILITIES IS TO GATHER FOREIGN INTELLIGENCE INFORMATION BY INTERCEPTING RADIO COMMUNICATIONS SENT TO OR FROM FOREIGN GOVERNMENTS.

THE AGENCY TOLD THE UFO BUFFS, "NSA MUST FOCUS ITS INTERCEPTION ACTIVITIES ON THOSE PARTICULAR COMMUNICATIONS LINES, CHANNELS, LINKS OR SYSTEMS WHICH YIELD THE HIGHEST PROPORTION OF USEFUL FOREIGN INTELLIGENCE INFORMATION.

"WHAT FOREIGN GOVERNMENTS DO NOT KNOW IS WHICH OF THE VAST NUMBER OF RADIO COMMUNICATIONS NSA ATTEMPTS TO INTERCEPT, WHICH ARE INTERCEPTED, AND, OF THOSE THAT ARE INTERCEPTED, WHICH YIELD TO NSA PROCESSING METHODS AND TECHNIQUES," NSA POLICY CHIEF EUGENE YEATES SAID IN A LETTER TO THE UFO GROUP.

"IT IS THE PROTECTION OF THIS CRITICAL INFORMATION THAT IS AT THE HEART OF THE INSTANT CASE," HE ADDED.

THAT REFUSAL PROMPTED THE ORGANIZATION TO FILE SUIT IN FEDERAL DISTRICT COURT IN WASHINGTON.

THE AGENCY, URGING THE COURT TO THROW OUT THE CASE, FILED ONE PUBLIC AFFIDAVIT AND ONE CONFIDENTIAL AFFIDAVIT THAT WAS REVIEWED IN PRIVATE BY JUDGE GERHARD GESELL.

THE JUDGE DISMISSED THE CASE IN 1980, DECLARING, "RELEASE OF THIS MATERIAL COULD SERIOUSLY JEOPARDIZE THE WORK OF THE AGENCY AND THE SECURITY OF THE UNITED STATES."

THE UFO GROUP THEN TOOK THE DISPUTE TO THE U.S. COURT OF APPEALS FOR THE DISTRICT OF COLUMBIA, BUT THAT PANEL AFFIRMED GESELL'S DECISION.

APPEALING TO THE SUPREME COURT, THE UFO BUFFS ARGUED, "THE DISTRICT COURT ACCEPTED WITHOUT QUESTION NSA'S OVERBROAD CLAIM OF EXEMPTION" FROM THE INFORMATION ACT.

THEY WENT ON TO ARGUE THAT UFOS HAVE "BEEN OBSERVED FOR THE PAST 35 YEARS BY RELIABLE AND RESPONSIBLE INDIVIDUALS INCLUDING SCIENTISTS AND MILITARY PERSONNEL."

UPI 02-15-82 07:38 PES

"Frontiers of Science" May/June 1981

8 East St S.E.
Wash DC 20003
phone 544-0233

What the U.S. Government Knows About Unidentified Flying Objects

by PETER GERSTEN

U.S. AIR FORCE

USAF Security intercepted a Cuban pilot's report of the encounter between his MIG-21 and a UFO.

At last! New evidence for the existence of unconventional aerial objects relies no longer on the credibility of civilian reports but on the records of scientists, military personnel, intelligence analysts, law enforcement officers and other reliable and responsible people. Their testimony can be found in three thousand pages of previously classified documents on UFOs released (mostly through Freedom of Information Act suits) over the past few years by the Departments of State/Army/Navy/Air Force/Defense, the Federal Bureau of Investigation, the National Security Agency, the Defense Intelligence Agency and the Central Intelligence Agency.

This overwhelming evidence indicates that Unidentified Flying Objects *do* exist, and that some of them are unconventional craft that (1) pose a threat to national security and (2) perform beyond the range of present-day technological development.

Furthermore, there is evidence that our government has continually misinformed the public concerning the true significance of the "UFO problem."

National Security and UFOs

"It is my view that this situation has possible implications for our national security."

—Central Intelligence Agency, 1952

In late 1952, a memorandum was drafted for CIA Director Walter B. Smith's signature, to be sent to the Executive Secretary of the National Security Council. The memo's subject: "Unidentified Flying Objects." The document shows that the CIA had "reviewed the current situation concerning unidentified flying objects which have caused extensive speculation in the press and has been the subject of concern to government organizations."

It was the Director's opinion, based has possible implications for our national security which transcend the interests of a single service.

"I therefore recommend that this Agency and the agencies of the Department of Defense be directed to formulate and carry out a program of intelligence and research activities required to solve the problem of instant positive identification of unidentified flying objects."

A draft of a proposed National Security Council directive was attached to the memorandum.

Unfortunately, it appears that the NSC directive fell by the wayside. Now, twenty-nine years later, the "current situation," contrary to official denials, still poses serious implications for our national security.

UFOs as a Threat

The Government's position:

"No UFO reported, investigated and evaluated by the Air Force has ever given any indication of a threat to our national security."

—Air Force, 1980

The evidence: reveal that during October, November, and December of 1975, reliable military personnel repeatedly sighted unconventional aerial objects in the vicinity of nuclear-weapons storage areas, aircraft alert areas and nuclear-missile control facilities at Loring Air Force Base, Maine; Wurtsmith AFB Michigan; Malstrom AFB, Montana; Minot AFB, North Dakota; and Canadian Air Forces Station, Ontario.[1] Many of the sightings were confirmed by radar. At Loring AFB, the interloper "demonstrated a clear intent on the weapons storage areas."

The incidents drew the attention of the CIA, the Joint Chiefs of Staff, and the Secretary of Defense. Though the Air Force informed the public and the press that individual sightings were isolated incidents, an Air Force document says that "Security Option III" was implemented and that security measures were coordinated with 15

New York attorney PETER A. GERSTEN has been pressing the legal effort on behalf of UFO groups—such as CAUS (Citizens Against UFO Secrecy)—for nearly three years. Gersten currently awaits a U.S. Appeals Court decision on release of over two hundred additional CIA documents relating to connection with the United States UFO program will be maintained in ASD" (plus, he added, a file of "finished intelligence reports").

Coinciding with the November 1957 UFO wave, "On 6 November 1957 a directed collection request for UFO information was levied on Contact Division by the Office of Scientific Intelligence... Responses to the requirement were forwarded to Collection Staff, OSI on 14 November 1957." (April 1, 1958 memo for Assistant Director, Scientific Intelligence, from Acting Assistant Director for Operations)

Also, coinciding with a flurry of sightings around the Washington, D.C. area in late 1964 and early 1965, another "collection request" was ordered by the Director of CIA. The present writer was among those interviewed by the Domestic Contact Division. The agent borrowed a number of specific UFO case investigation reports.

A series of memos from April through late summer of 1976 shows renewed interest in UFO data, probably as a result of the 1975 and 1976 UFO sightings at Strategic Air Command bases and other military installations. One censored message of April 3, 1976, refers to "Subject: New DCD Case [censored]—UFO Research. Reference: Form 610 dated 9 April 1976 transmitting UFO Study." Date discrepancy notwithstanding, the memos go on to discuss "the UFO study." One, while stating that there was at the time no formal UFO program, read: "At the present time, there are offices and personnel within the Agency who are monitoring the UFO phenomena... not currently on an official basis. Dr. [censored] feels that the best approach would be to keep in touch with and in fact develop reporting channels in this area to keep the Agency/community informed of any new developments."

Where are the startling 1952 cases the led the CIA to the brink of a major scientific study?

Memos ranging from 1958 to 1967 mention analyses of UFO photos by the CIA National Photographic Interpretation Center under Arthur C. Lundahl, partially confirming information from private sources. Indeed, Lundahl was involved in arranging a "window" on the Colorado Project and in secret meetings with Dr. Edward Condon and members of the Colorado Project staff.

Glaringly absent from the released documents are the photo analyses, including one case provided to a scientific firm by the present writer that found its way to the CIA, and the hundreds and hundreds of intelligence reports on individual UFO cases that must have been collected—and in some cases are known to have been collected. Also absent are the dozens of radar-visual cases which must have been studied by the Physics and Electronics Division, as well as other materials that made up the case files repeatedly alluded to in these documents. Where are the startling 1952 cases that led the CIA to the brink of a major scientific study of UFOs?

The 892 pages contain large amounts of trivial correspondence, administrative papers, duplication and irrelevant (non-UFO) documents. Obviously, they do *not* contain the CIA UFO case files, other than a random sprinkling of mostly known cases. Only the FOIA lawsuits have been able to pry loose a few highly significant cases, such as the 1976 Iranian jet case.

There can be no doubt that the CIA is stonewalling. If *this* is all the case files developed in more than thirty years, I would have to conclude that the CIA is totally inept. And, of course, it isn't. ☐

In a Just CAUSe

Over the past third of a century, the government's conduct with regard to UFOs has been characterized as nonfeasance, misfeasance, and malfeasance. Citizens Against UFO Secrecy, a public-interest group, was formed to foster a review of the reality and significance of UFOs and the government's policies and practices regarding them.

CAUS calls upon the Federal government to (1) admit that the public has been misled about the nature of UFOs, (2) acknowledge that UFOs exist, and (3) reverse its position that further scientific study of UFO reports is unwarranted. CAUS seeks the immediate declassification and public dissemination of all official UFO documentation.

CAUS believes that the public has a right to an objective reappraisal of the implications of the UFO phenomenon.

For more information, write to CAUS, P.O. Box 4743F, Arlington, Virginia 22204.

FOOTNOTES

1. NORAD Classified Message; to Secretary of Air Force, *et al;* from NORAD Director of Operations; subject: Suspicious Unknown Air Activity; November 11, 1975. (Still classified CONFIDENTIAL; leaked to UFO research group in late 1976.)
2. Air Force Security Police (AFSP) message to 15 Air Force Bases, November 10, 1975.
3. This and the previous statements are found in the following documents:
   - Memorandum for Record; for the Joint Staff; signed by J.B. Morin, Rear Admiral, USN & Deputy Director for Operations, NMCC: January 21, 1976.
   - Memorandum for Record; for the Joint Staff; from Fred A. Treyz, Brigadier General, USAF & Deputy Director for Operations, NMCC: subject: Unidentified Flying Object Sighting: January 31, 1976.
   - Memorandum for Record; for the Joint Staff; from L.J. LeBlanc, Jr., Brigadier General, USMC & Deputy Director for Operations, NMCC: subject: Reports of Unidentified Flying Objects (UFOs); July 30, 1976.
4. "Rees Letter": AFOSI (Air Force Office of Special Investigations), 17th District, Kirtland AFB, New Mexico; letter to director of AFOSI; May 25, 1950.
5. "Cuban Incident": Air Force Security Service 6947th Squadron technician's statement to Stanton T. Friedman; October 27, 1977.
6. Memorandum for Director of Central Intelligence; subject: Flying Saucers; by E. Marshall Chadwell, Assistant Director, Scientific Intelligence; September 24, 1952.
7. Memo; to E. James Archer and Thurston E. Hanning; from Robert J. Low; subject: "Some Thoughts on the UFO Project": August 9, 1966.

The Government and UFOs

"Further scientific investigation of UFOs is unwarranted."

—Air Force, 1980

Perhaps most disturbing is the very fact that after thirty-two years, a small but significant percentage of UFOs still remains unidentified. While the government has been concerned with the psychological danger the UFO phenomenon poses, it has been unwilling to consider the prospect that some UFOs pose an actual physical threat. Fearful of generating undue concern, the government has deliberately chosen to debunk UFO reports and has misinformed the public as to the true importance of the phenomenon.

Unconventional aerial objects that boast unlimited and unrestricted access to our most sensitive nuclear installations—and which can render inoperable the instrumentation, communication/weapon systems of American-made jets, or which can shut down and restart at will sophisticated hydraulic equipment—*do* warrant further scientific study. Awareness of an advanced technology and potential threat is not an unreasonable pursuit. As the National Security Agency indicates, it could be a matter of survival.

Though admittedly the government has studied UFO reports, apparently no government body has dwelt on those official government reports that indicate certain UFOs pose a threat to national security. Is there any doubt that it is these reports which deserve further scientific investigation?

The now-defunct USAF twenty-year "Project Blue Book" UFO study never had a chance to receive the "outstanding report" from Iran. An Air Force document states: "Reports of UFOs which could affect national security are made in accordance with JANAP 146* or Air Force Manual 55-11, and are not part of the Blue Book system." The Air Force's UFO investigation was criticized as long ago as 1952 by the CIA. The CIA complained that the Air Force's case-by-case investigations and explanations were insufficient to determine the exact nature of the phenomenon.[6]

Similarly, the Air Force-sponsored "Condon Committee" study by the University of Colorado in 1968 never earnestly intended to investigate the physical reality of the phenomenon. Indeed, an early memorandum by one of Dr. Edward U. Condon's staff indicates otherwise: "The trick would be, I think to describe the project so that to the public, it would appear a totally objective study . . . one way to do this would be to stress investigation, not of the physical phenomenon, but rather of the people who are doing the observing . . ."[7]

Conclusion

In June 1978, a French government UFO study group (GEPAN) concluded that "everything taken into consideration, a material phenomenon seems to be behind the totality of the phenomenon—a flying machine whose modes of sustenance and propulsion are beyond our knowledge."

If the UFO phenomenon is indeed beyond the grasp of our understanding—technologically speaking—all the more reason to strive towards learning more about it. For although the United States may ignore the significance of the UFO phenomenon, it is hardly reasonable to suppose that the rest of the world will do so.

And there are other considerations besides national security in following up the UFO enigma. As a report from the National Security Agency in 1968 put it,

> "Perhaps the UFO question might even make man undertake studies which could enable him to construct a society which is most conducive to developing a completely human being, healthy in all aspects of mind and body—and, most importantly, able to recognize and adapt to real environmental situations."

?
?
?

In isolating ourselves from the UFO phenomenon we may risk missing what could be the most important adventure man has yet embarked upon. □

Joint Army-Navy-Air Force Publication 146 is published by the Military Communications Board of the DOD Joint Chiefs of Staff. It provided U.S.-Canadian "Communications Instructions for Reporting Vital Intelligence Sightings (CIRVIS) from Airborne and Waterborne Sources." Section III (Security), paragraph 208, calls for stiff penalties for divulging information about such sightings once reported. —Ed.

Is the CIA Stonewalling?

by RICHARD HALL

BASED ON the 892 pages of UFO-related documents released to lawyer Peter Gersten, it is clear that the CIA's professed non-interest in UFOs is untrue. There is internal evidence of non-continuity within the CIA, and even of one group or analyst being unaware of other files or previous work. This is not surprising considering the highly compartmentalized nature of the agency. But periodic studies or reviews were ordered and UFO reports, foreign and domestic, were routinely monitored over long periods of time.

The statement is sometimes made that the CIA has had no "formal" study of UFOs other than the 1953 Robertson Panel, but these documents show that the agency kept (and no doubt keeps) plenty of "channels" open to gather information, including an acknowledged channel into the 1966-68 University of Colorado UFO project. The documents also clearly indicate that in 1952 the CIA was prepared to mount a major scientific study of UFOs based on the extraordinary radar-visual sightings that year, but that the debunking conclusion of the Robertson Panel cut short that effort.

In May 1953, following the Robertson Panel report, "P&E Division assumed responsibility for the OSI project on unidentified flying objects." (December 17, 1953 memo to Assistant Director, Scientific Intelligence) A year later, the Chief of P&E (Physics and Electronics) said he would "maintain the OSI central file on such objects," which he did until August 1955, according to an August 8, 1955 memo to the Assistant Director of Scientific Intelligence.

AFTER THAT, responsibility was assigned to the Applied Science Division, where W.E. Lexow, Division Chief, stated in a February 1956 memo: "A chronological file of all OSI correspondence and action taken in

RICHARD HALL was former Assistant Director of NICAP, the leading UFO organization during the 1950s and 1960s. He is the Editor of MUFON Journal *and a* Frontiers of Science *advisor.* □

[redacted]

NSA ↓

Air Force bases from Guam to w-foundland.[2] Another AF docu...ent reveals that the Air Force conducted an investigation into the incidents but found no explanation for their occurrence.

It appears Air Force "security measures" provided no protection against the "invasion." One month later, on January 21, 1976, UFOs "25 yards in diameter, gold or silver in color with blue light on top, hole in middle, and red light on bottom" were observed "near the flight line of Cannon AFB, N.M." Ten days later, on January 31, a UFO was observed near a radar site at Elgin AFB, Florida. On July 30, 1976, a UFO was observed "over the ammo storage area" at Fort Richie, Maryland.[3]

The above accounts have numerous historical precedents. From 1948 through 1950, an FBI document reveals, UFOs were sighted by "persons whose reliability is not questioned," near highly sensitive military and government installations, including nuclear weapons design, construction, testing and stockpiling sites. Security officials were greatly alarmed by these incidents.[4]

A CIA document reveals that in 1952 "sightings of unexplained objects at great altitudes and travelling at high speeds" were reported in the vicinity of major U.S. defense installations and posed a threat to national security.

The evidence is clear and convincing that the Federal government has systematically misinformed the American people about the real threat to our national security posed by such UFO encounters.

UFO As Advanced Technology

The Government's position:

"There has been no evidence submitted to or discovered by the Air Force that sightings categorized as 'unidentified' represent technological developments or principles beyond the range of present-day scientific knowledge."

– Air Force, 1980

The official documents reveal hundreds of sighting reports – many confirmed by radar and other tracking devices – that describe unconventional objects exhibiting advanced performance characteristics involving maneuverability, speed, size a shape.

A Defense Intelligence Agency document reveals that on September 19, 1976, American-made Iranian jets encountered several UFOs that exhibited a technology beyond present-day development. During the night-time encounter, one F-4 jet, upon approaching one of the UFOs, lost all instrumentation and communications functions. Another F-4's weapons-control panel became inoperable when the pilot attempted to fire at the object.

The Federal Government has systematically misinformed the American people about the real threat...

The DIA evaluation (October 12, 1976) refers to this incident as "an outstanding report" because the objects were seen by many witnesses of high credibility; the visual sightings had radar confirmation; similar electromagnetic effects were reported by three separate aircraft; and physiological effects were reported by some of the crew members. Furthermore, the UFOs displayed an "inordinate amount of maneuverability."

A State Department message (March 7, 1975) from the American Embassy in Algiers tells of "strange machines" seen near Algerian military installations by "respectable people." Some of the sightings were confirmed by radar.

And another State Department message from our embassy in Kuwait reports that during November 1978, a series of UFO sightings caused the Kuwaiti government to appoint an investigatory committee from the Kuwait Institute for Scientific Research. One UFO appearing over the northern oil fields "seemingly did strange things" to the automatic pumping equipment. The equipment is designed to shut itself down when any failure occurs that could seriously damage the petroleum-gathering and transmission system; when such an event occurs, the pumping equipment must be restarted manually. When the UFO appeared, the pumping system automatically shut down. But when the UFO "vanished," the system started up again, *automatically.*

The presence of a highly sophisticated technology – a technology beyond our present development – seems obvious. Why is it being ignored by our government?

A Question of Survival

"It would seem a little more of this survival attitude is called for in dealing with the UFO problem."

– National Security Agency, 1968

The evidence indicates that some unconventional aerial objects could provoke, either intentionally or unintentionally, an international incident – with serious repercussions.

In March 1967, an intercept technician with the USAF Security Service intercepted a communication between the pilot of a Russian-made Cuban MIG-21 and his command concerning a UFO encounter.[5] The technician has since stated that when the pilot attempted to fire at the object, the MIG and its pilot were destroyed by the UFO. Furthermore, the technician alleges that all reports, tapes, log entries, and notes on the incident were forwarded to the National Security Agency at their request.

Not surprisingly, several months later, the agency drafted a report entitled *UFO Hypothesis and Survival Question.* Released in October 1979 under the U.S. Freedom of Information Act, the report states that "the leisurely scientific approach has too often taken precedence in dealing with the UFO question." The Agency concluded that no matter what UFO hypothesis is considered, "all of them have serious survival implications."

Comparing the UFO problem to a rattler on a forest path, the NSA report says, "you would have to treat the alarm as if it were a real and immediate threat to your survival. Investigation would become an intensive emergency action to isolate the threat and to determine its precise nature. It would be geared to developing adequate defense measures in a minimum amount of time. It would seem a little more of this survival attitude is called for in dealing with the UFO problem."

FRENCH GOVERNMENT UFO STUDY

"Presentation to the Scientific Counsel of G.E.P.A.N. of Studies Undertaken During the First Semester of 1978," (June 1978, 5 volumes, approximately 500 pages).

This report of the French governmental UFO study group GEPAN[1] documents the studies carried out by the group during the first part of 1978. Three special groups (rapid intervention, physical traces, radar alert) were created as part of the study group's functions, but they were little used during 1978. Instead, the bulk of work was devoted to 11 cases of high credibility and high strangeness. Eleven such cases were studied in great detail; only one proved to have a conventional explanation. In the other 10, it appeared that the distance between the witnesses and the objects was less than 250 meters. Of the five volumes of the report, three were entirely devoted to analysis of these 11 cases, all except one of which was pre-1978. The earliest was 1966. Two of the cases were humanoid sightings.

The analysis and investigation was carried out by a four-person team in each case; the team included a psychologist, who separately carried out a psychological examination relevant to the evaluation of the testimony of the witnesses. The care with which distances, angles, and psychological factors were evaluated makes the bulk of the Condon Report seem very poor by comparison. In many cases, the investigations were textbook models of how such investigations should be carried out.

In 10 of the 11 cases, the conclusion was that the witnesses had witnessed a material phenomenon that could not be explained as a natural phenomenon or a human device. One of the conclusions of the total report is that behind the overall phenomenon there is a "flying machine. . . whose modes of sustenance and propulsion are beyond our knowledge."

GEPAN was created in 1977, under the direction of Dr. Claude Poher. Poher has now resigned, feeling that he has done everything he could do with the methods at hand. Nonetheless, the organization has not been disbanded but continues its work. During 1977, it was largely concerned with checking Poher's statistics, which it approved.[2] During 1978, it has expanded its operation into the investigation of actual cases. It is alerted to actual cases by teletype by the Gendarmerie, the French national police force. The above report has reportedly been approved by GEPAN's supervisory scientific counsel.

Note: the original report was limited to 120 or 140 copies and was secret. It is not available for general dissemination, and in any case is in French. There is extensive coverage in the report of the Teheran case, but nothing beyond what is known to American researchers; it was not one of ~~the~~ eleven cases studied, all of which were in France.

---

[1]Group d'Etudes Des Phenomenes Aerospatiaux Non-Identifiees

[2]Although it did study two nocturnal light cases, both of which ended up labelled "unidentified". (Submitted by Ron Westrum.)

This page intentionally left blank.

Communications Intelligence (COMINT) Report - "XXXXX Unidentified Flying Objects"

TOP SECRET Documents Redacted and Released by the NSA as a Result of Civil Action No. 80-1562

This page intentionally left blank.

[redacted]

[redacted] UNIDENTIFIED FLYING OBJECTS
[redacted]

REQS: [redacted]

TEXT:

[redacted] UNIDENTIFIED FLYING OBJECTS
(UFO) ON [redacted] AWARE OF
VARIOUS UNIDENTIFIED OBJECTS IN [redacted]

[redacted] AN
UNIDENTIFIED SILENT LIGHT MOVING [redacted] THE LIGHT WAS A SATELLITE NOT AN AIRCRAFT [redacted] AN
UNIDENTIFIED LIGHT [redacted]
THE LIGHT WAS SUBSEQUENTLY IDENTIFIED AS AT LEAST ONE AIRCRAFT. [redacted]

THREE STRANGE LIGHTS (NFI) [redacted] ONE WAS A STATIONARY, BLINKING
LIGHT; THE TWO OTHER, MOVING, LIGHTS CROSSED PATHS. [redacted] THE UFO WAS AT AN ALTITUDE OF APPROXIMATELY 300 [redacted]
METERS [redacted] AIRCRAFT
IN THE AREA. [redacted]

[redacted]

[redacted]

[redacted]

[redacted] IT WAS A CARGO FLIGHT.]STET

ADMIN

[redacted]

[redacted]

[redacted] ~~SECRET~~

COMNAVSECGRU
Visual Sighting Report

This page intentionally left blank.

COMNAVSECGI N APRIL 30, 1979

AT 220032 AUG VIS OBSVD BRIGHT WHITE LIGHT APPEARING

AT BEARING 026 (TRUE NORTH) ELEVATION APPROX 15 DEGREES.

23 17 82-19W

POSIT OF MULLER [illegible] DURING THE FIRST MIN

OF OBSERVATION THE LIGHT WAS PULSATING OR FLUCTUATING

IN BRIGHTNESS AT APPROX ONE SECOND INTERVALS. FOR THE

NEXT TWO MINS OF OBSERVATION THE LIGHT BEGAN PULSATING

[illegible] CONCENTRIC RINGS OF LIGHT WHICH FLOWED FROM THE CENTRAL

(S)OURCE IN A RIPPLING, SHIMMERING FASHION DIMINISHING IN

BRIGHTNESS AS THEY GREW FURTHER AND FURTHER AWAY FROM

THE ORIGINATING SOURCE. DURING THE FOURTH MIN OF

OBSERVATION A BLUISH-GREEN BEAM OF LIGHT APPEARED FROM

THE CENTRAL CORE OF THE CONFIGURATION, EXTENDING OUTBOARD

AND DOWNWARD TO THE LEFT AT AN ANGLE OF APPROX. 45 DEGREES,

AND REACHING TO THE FADE OUT POINT OF THE RADIATING RINGS

OF LIGHT. APPROX FIVE MINS AFTER THE APPEARANCE OF THE

BLUISH-GREEN BEAM (SIMILAR IN APPEARANCE TO A SEARCHLIGHT

BEAM), THE RADIATING CIRCLES OF LIGHT DISAPPEARED, LEAVING

ONLY THE ORIGINAL W[illegible] CORE OF LIGHT AND THE [illegible] LORED BEAM. FOLLOWING THE DISAPPEARANCE OF THE CIRCLES, THE BEAM SEEMED TO RETRACT TOWARDS THE CENTRAL CORE, GROWING BROADER AND STOUTER AS IT RETRACTED. FINALLY, THE BEAM AND CORE SEEMED TO FORM INTO A NEBULOUS CLOUD-SHAPED MASS, WHICH DRIFTED SOUTH ON THE WIND, FADING FROM VIEW AFTER APPROX 14 MINS. DURING THIS EVENT, THE FOLLOWING ATMOSPHERIC DATA WAS ACCUMULATED: BAROMETER 29.91. TEMPERATURE - 83 DEGREES DRY/78 DEGREES WET. VISIBILITY EXCELLENT, FOUR-TENTHS CLOUD COVER (STRATUS ~~AND~~ CUMULUS). WIND E X N.S. AT 12KTS, ELECTRICAL STORM FORMING OVER IMMEDIATE AREA OF CITY OF HAVA.

THIS DOCUMENT DECLASSIFIED BY
COMNAVSECGRU ON APRIL 30, 1979

Report Bibliography

Issued by the Defense Documentation Center

This page intentionally left blank.

REPORT BIBLIOGRAPHY

3370

Defense Documentation Center

Defense Supply Agency

Cameron Station • Alexandria, Virginia 22314

This Report Bibliography has been prepared by the Defense Documentation Center as a mission assignment in accordance with the provisions of DoD Instruction 5100.38.

ROBERT B. STEGMAIER, JR.
Administrator
Defense Documentation Center

UNCLASSIFIED

NOTICE

WHEN GOVERNMENT OR OTHER DRAWINGS, SPECIFICATIONS OR OTHER DATA ARE USED FOR ANY PURPOSE OTHER THAN IN CONNECTION WITH A DEFINITELY RELATED GOVERNMENT PROCUREMENT OPERATION, THE U.S. GOVERNMENT THEREBY INCURS NO RESPONSIBILITY, NOR ANY OBLIGATION WHATSOEVER; AND THE FACT THAT THE GOVERNMENT MAY HAVE FORMULATED, FURNISHED, OR IN ANY WAY SUPPLIED THE SAID DRAWINGS, SPECIFICATIONS, OR OTHER DATA IS NOT TO BE REGARDED BY IMPLICATION OR OTHERWISE AS IN ANY MANNER LICENSING THE HOLDER OR ANY OTHER PERSON OR CORPORATION, OR CONVEYING ANY RIGHTS OR PERMISSION TO MANUFACTURE, USE OR SELL ANY PATENTED INVENTION THAT MAY IN ANY WAY BE RELATED THERETO.

LIMITED REPORTS

REFERENCES TO ANY REPORTS LIMITED IN DISTRIBUTION ARE INCLUDED IN THIS BIBLIOGRAPHY FOR REFERENCE PURPOSES ONLY. TO OBTAIN COPIES OF THESE REPORTS, REQUESTS SHOULD BE FORWARDED TO THE CONTROLLING AGENCY VIA THE PROJECT OFFICER RESPONSIBLE FOR YOUR CONTRACT. SUCH REQUESTS SHOULD INCLUDE ALL DESCRIPTIVE CATALOGING INFORMATION NECESSARY FOR ACCURATE IDENTIFICATION.

NOFORN OR SIMILIAR MARKINGS

THE ENTRY SO MARKED IS SUBJECT TO SPECIAL EXPORT CONTROLS AND EACH TRANSMITTAL TO A FOREIGN GOVERNMENT OR FOREIGN NATIONAL MAY BE MADE ONLY WITH PRIOR APPROVAL OF THE ACTIVITY CITED IN THE BIBLIOGRAPHIC ENTRY.

NON-PERTINENT REFERENCES

ALL DDC BIBLIOGRAPHIES ARE PRODUCED BY A COMPUTER SEARCH OF OUR DATA BANK. THESE BIBLIOGRAPHIES MAY OR MAY NOT HAVE BEEN REVIEWED BY A TECHNICAL SPECIALIST. IN THE EVENT A REVIEW IS MADE AND NON PERTINENT REFERENCES ARE FOUND, THEY MAY OR MAY NOT HAVE BEEN REMOVED FROM THE BIBLIOGRAPHY. IF NON-PERTINENT REFERENCES ARE RETAINED IN A REVIEWED BIBLIOGRAPHY, THEY WILL BE STAMPED NON-PERTINENT . BLANK PAGES ARE OCCASIONALLY INCLUDED IN BIBLIOGRAPHIES. THESE PAGES ARE NOT THE RESULT OF COMPUTER MALFUNCTIONS; THEY ARE THE RESULT OF ASSEMBLY PROCEDURES, WHICH ARE DESIGNED TO EXPEDITE OUR SERVICE TO YOU.

COMPLAINTS

IF YOU RECEIVE A BIBLIOGRAPHY THAT DOES NOT MEET YOUR REQUIREMENTS, PLEASE REPORT IT TO THE CHIEF OF THE BIBLIOGRAPHY BRANCH BY CALLING 202 - 694-7058. PLEASE CITE THE SEARCH CONTROL NUMBER OF THE BIBLIOGRAPHY WHEN YOU CALL. EVERY EFFORT WILL BE MADE TO PROVIDE YOU THE INFORMATION THAT YOU NEED.

UNCLASSIFIED

DDC REPORT BIBLIOGRAPHY SEARCH CONTROL NO. 031302

AD-688 541 1/2 22/1
NATIONAL ACADEMY OF SCIENCES WASHINGTON D C
REVIEW OF THE UNIVERSITY OF COLORADO REPORT ON UNIDENTIFIED FLYING OBJECTS BY A PANEL OF THE NATIONAL ACADEMY OF SCIENCES. (U)
DESCRIPTIVE NOTE: SPECIAL REPT.,
JAN 69 10P CLEMENCE,GERALD M. ;
CONTRACT: F18600-67-C-0071
PROJ: AF-9730
MONITOR: AFOSR 69-1276TR

UNCLASSIFIED REPORT
PORTIONS OF THIS DOCUMENT ARE ILLEGIBLE. SEE INTRODUCTION SECTION OF THIS ANNOUNCEMENT JOURNAL FOR CFSTI ORDERING INSTRUCTIONS.

DESCRIPTORS: (*OPTICAL PHENOMENA, REVIEWS), VISUAL PERCEPTION, REPORTS (U)
IDENTIFIERS: *UNIDENTIFIED FLYING OBJECTS (U)

THE AD HOC COMMITTEE CONCURS WITH THE FINDINGS OF THE REPORT OF THE UNIVERSITY OF COLORADO ENTITLED 'SCIENTIFIC STUDY OF UNIDENTIFIED FLYING OBJECTS,' DR. EDWARD U. CONDON, SCIENTIFIC DIRECTOR (SEE AD-680 975, AD-680 976, AND AD-680 977). PANEL AGREES THAT A STUDY OF UFOS IN GENERAL IS NOT A PROMISING WAY TO EXPAND SCIENTIFIC UNDERSTANDING OF THE PHENOMENA, AND ADDS: 'ON THE BASIS OF PRESENT KNOWLEDGE THE LEAST LIKELY EXPLANATION OF UFOS IS THE HYPOTHESIS OF EXTRATERRESTRIAL VISITATIONS BY INTELLIGENT BEINGS.' (AUTHOR) (U)

DDC REPORT BIBLIOGRAPHY SEARCH CONTROL NO. 031302

AD-687 960 6/5
COLORADO UNIV BOULDER
A CASE OF 'AUTOSTASIS' OR REVERSE
AUTOKINESIS, (U)
FEB 68 3P WERTHEIMER, MICHAEL ;
CONTRACT: F44620-67-C-0035
PROJ: AF-9730
MONITOR: AFOSR 69-1150TR

UNCLASSIFIED REPORT
AVAILABILITY: PUB. IN PERCEPTUAL AND MOTOR
SKILLS, V26 P417-418 1968.

DESCRIPTORS: (*VISUAL PERCEPTION, ILLUSIONS),
(*ILLUSIONS, *OPTICAL PHENOMENA), PHYSIOLOGY,
VISION, HUMANS (U)
IDENTIFIERS: AUTOKINESIS, AUTOSTASIS,
UNIDENTIFIED FLYING OBJECTS (U)

THREE OF FIVE OBSERVERS OF A LIGHT IN THE NIGHT SKY THAT WAS ACTUALLY MOVING CONTINUOUSLY ALONG A LINEAR COURSE REPORTED IT AS STATIONARY AS LONG AS THE LIGHT WAS ON. THIS PHENOMENON, 'AUTOSTASIS,' SEEMS TO BE OPPOSITE TO THE WELL-KNOWN PHENOMENON OF AUTOKINESIS, OR APPARENT MOTION OF AN ACTUALLY STATIONARY LIGHT IN ~~AN~~ UNDIFFERENTIATED FIELD. (AUTHOR) (U)

DDC REPORT BIBLIOGRAPHY SEARCH CONTROL NO. 031302

AD-680 976 1/2 22/1 5/10
COLORADO UNIV BOULDER
SCIENTIFIC STUDY OF UNIDENTIFIED FLYING OBJECTS,
VOLUME 2. (U)
DESCRIPTIVE NOTE: FINAL REPT.,
JAN 69 419P CONDON,EDWARD U. ;
CONTRACT: F44620-67-C-0035
PROJ: AF-9730
MONITOR: AFOSR 69-0026TR

UNCLASSIFIED REPORT

SUPPLEMENTARY NOTE: SEE ALSO VOLUME 1, AD-680 975 AND
VOLUME 3, AD-680 977.

DESCRIPTORS: (*OPTICAL PHENOMENA, AIR FORCE
RESEARCH), PHOTOGRAPHIC ANALYSIS, OPTICAL
ANALYSIS, HISTORY, VISUAL PERCEPTION (U)
IDENTIFIERS: *UNIDENTIFIED FLYING OBJECTS, CASE
STUDIES (U)

THE REPORT CONTAINS THE RESULTS OF A SCIENTIFIC
INQUIRY INTO THE PHENOMENA OF UNIDENTIFIED FLYING
OBJECTS. THIS VOLUME CONTAINS: (1) CASE
STUDIES DURING THE TERM OF THE CONTRACT, (2)
PHOTOGRAPHIC CASE-STUDIES, AND (3) HISTORICAL
ASPECTS OF UFO PHENOMENA. (AUTHOR) (U)

DDC REPORT BIBLIOGRAPHY SEARCH CONTROL NO. 031302

AD-688 332 1/2 22/1
COLORADO UNIV BOULDER
UFOS AND RELATED SUBJECTS: AN ANNOTATED
BIBLIOGRAPHY, (U)
DESCRIPTIVE NOTE: SPECIAL REPT.,
JUL 69 415P CATOE, LYNN E. ;
CONTRACT: F44620-67-C-0035
PROJ: AF-9730
MONITOR: AFOSR 68-1656

UNCLASSIFIED REPORT
AVAILABILITY: PAPER COPY AVAILABLE FROM
SUPERINTENDENT OF DOCUMENTS, GPO, WASHINGTON, D.
C. 20402, $3.50.
SUPPLEMENTARY NOTE: PREPARED IN COOPERATION WITH LIBRARY
OF CONGRESS, WASHINGTON, D. C. SCIENCE AND
TECHNOLOGY DIV. LC-68-62196.

DESCRIPTORS: (*OPTICAL PHENOMENA,
*BIBLIOGRAPHIES), BIOLOGY, SOLAR SYSTEMS,
LIGHTNING, AIRCRAFT, DISKS, RELIGION,
GRAVITY, ILLUSIONS, THEORY, ABSTRACTS (U)
IDENTIFIERS: *UNIDENTIFIED FLYING OBJECTS (U)

THE REPORT IS AN ANNOTATED BIBLIOGRAPHY OF BOOKS,
JOURNAL ARTICLES, PAMPHLETS, CONFERENCE PROCEEDINGS,
TAPES, ORIGINAL MANUSCRIPTS, BOOKS OF PHOTOGRAPHS,
CARTOONS, MOTION PICTURE FILMS AND OTHER SUCH
MATERIAL ON THE SUBJECT OF UNIDENTIFIED FLYING
OBJECTS (UFOS), TOTALING MORE THAN 1,600 SEPARATE
ITEMS, WITH AUTHOR INDEX. MAJOR CATEGORIES
INCLUDE: UFOS, ORIGIN OF LIFE, MANKIND, SOLAR
SYSTEM, EXTRATERRESTRIAL LIFE, E.T. VISITORS, BALL
LIGHTNING AND FIREBALLS, DISC-LIKE AIRCRAFT,
UNIDENTIFIED SUBMARINE OBJECTS, FORTEAN PHENOMENA,
UFOS AND RELIGION, UFOS AND TIME, GRAVITY AND
ANTI-GRAVITY, HOLLOW EARTH THEORY, DISAPPEARANCES,
CARTOONS, MIRAGES, AND RELATED SUBJECTS.
(AUTHOR) (U)

UNCLASSIFIED

DDC REPORT BIBLIOGRAPHY   SEARCH CONTROL NO. 031302

AD-680 977   1/2   22/1   5/10
COLORADO UNIV BOULDER
SCIENTIFIC STUDY OF UNIDENTIFIED FLYING OBJECTS,
VOLUME 3. (U)
DESCRIPTIVE NOTE: FINAL REPT.,
JAN 69   527P   CONDON,EDWARD U. ;
CONTRACT: F44620-67-C-0035
PROJ: AF-9730
MONITOR: AFOSR 69-0027TR-3

UNCLASSIFIED REPORT

SUPPLEMENTARY NOTE: SEE ALSO VOLUME 2, AD-680 976.

DESCRIPTORS: (*OPTICAL PHENOMENA, AIR FORCE RESEARCH), VISUAL PERCEPTION, PSYCHOLOGY, RADAR, SONIC BOOM, ATMOSPHERIC ELECTRICITY, PLASMA MEDIUM, BALLOONS, STATISTICAL ANALYSIS, PUBLIC OPINION, ILLUSIONS (U)
IDENTIFIERS: *UNIDENTIFIED FLYING OBJECTS, CASE STUDIES (U)

THE REPORT CONTAINS THE RESULTS OF A SCIENTIFIC INQUIRY INTO THE PHENOMENA OF UNIDENTIFIED FLYING OBJECTS. THIS VOLUME CONTAINS: (1) THE SCIENTIFIC CONTEXT, (2) APPENDICES A-X, AND (3) THE INDEX. (AUTHOR) (U)

DDC REPORT BIBLIOGRAPHY SEARCH CONTROL NO. 031302

AD-680 975 1/2 22/1 5/10
COLORADO UNIV BOULDER
SCIENTIFIC STUDY OF UNIDENTIFIED FLYING OBJECTS,
VOLUME 1. (U)
DESCRIPTIVE NOTE: FINAL REPT.,
JAN 69 443P CONDON,EDWARD U. ;
CONTRACT: F44620-67-C-0035
PROJ: AF-9730
MONITOR: AFOSR 69-0025TR

UNCLASSIFIED REPORT

SUPPLEMENTARY NOTE: SEE ALSO VOLUME 2, AD-680 976.

DESCRIPTORS: (*OPTICAL PHENOMENA, AIR FORCE RESEARCH), PHOTOGRAPHIC ANALYSIS, OPTICAL ANALYSIS, RADAR, ASTRONAUTS, ATTITUDES, HISTORY, VISUAL PERCEPTION, SONIC BOOM, ATMOSPHERIC ELECTRICITY, PLASMA MEDIUM, BALLOONS, STATISTICAL ANALYSIS, PUBLIC OPINION, ILLUSIONS (U)
IDENTIFIERS: *UNIDENTIFIED FLYING OBJECTS, CASE STUDIES (U)

THE REPORT CONTAINS THE RESULTS OF A SCIENTIFIC INQUIRY INTO THE PHENOMENA OF UNIDENTIFIED FLYING OBJECTS. THIS VOLUME CONTAINS: (1) CONCLUSIONS AND RECOMMENDATIONS, (2) SUMMARY OF THE STUDY, AND (3) CASE STUDIES PREDATING THE TERM OF THE PROJECT. (AUTHOR) (U)

This page intentionally left blank.

Now You See It, Now You Don't

by Captain Henry S. Shields,
HQ USAFE/INOMP

This page intentionally left blank.

NOW YOU SEE IT, NOW YOU DON'T! ( )

Captain Henry S. Shields, HQ USAFE/INOMP

[redacted] Sometime in his career, each pilot can expect to encounter strange, unusual happenings which will never be adequately or entirely explained by logic or subsequent investigation. The following article recounts just such an episode as reported by two F-4 Phantom crews of the Imperial Iranian Air Force during late 1976. No additional information or explanation of the strange events has been forthcoming; the story will be filed away and probably forgotten, but it makes interesting, and possibly disturbing, reading.

* * * * *

[redacted] Until 0030 on a clear autumn morning, it had been an entirely routine night watch for the Imperial Iranian Air Force's command post in the Tehran area. In quick succession, four calls arrived from one of the city's suburbs reporting a series of strange airborne objects. These Unidentified Flying Objects (UFOs) were described as 'bird-like', or as brightly-lit helicopters (although none were airborne at the time). Unable to convince the callers that they were only seeing stars, a senior officer went outside to see for himself. Observing an object to the north like a star, only larger and brighter, he immediately scrambled an IIAF F-4 to investigate.

[redacted] Approaching the city, the F-4 pilot reported that the brilliant object was easily visible 70 miles away. When approximately 25 NM distant, the interceptor lost all instrumentation and UHF/Intercom communications. Upon breaking off the intercept and turning towards his home base, all systems returned to normal, as if the strange object no longer regarded the aircraft as a threat.

DECLASSIFY ON: 4 Dec 81
by: ACS/I HQUSAF

[redacted] A second F-4 was scrambled ten minutes after the first. The backseater reported radar-lock on the UFO at 27 NM/12 o'clock high position, and a rate of closure of 150 knots. Upon reaching the 25 NM point, the object began rapidly moving away to maintain a constant separation distance while still visible on the radar scope. While the size of the radar return was comparable to that of a KC-135, its intense brilliance made estimation of actual size impossible. Visually, it resembled flashing strobe lights arranged in a rectangular pattern and alternating blue, green, red, and orange. Their sequence was so fast that all colors could be seen at once.

[redacted] As the F-4 continued pursuit south of Tehran, a second brightly-lit object (about one-half to one-third the size of the moon) detached from the original UFO and headed straight for the F-4 at a high rate of speed. The pilot attempted to fire an AIM-9 missile at the new object but was prevented by a sudden power loss in his weapons control panel. UHF and internal communications were simultaneously lost. The pilot promptly initiated a turn and negative-G dive to escape, but the object fell in behind the F-4 at 3-4 NM distance. Continuing the turn, the pilot observed the second object turn inside of him and then away, subsequently returning to the primary UFO for a perfect rendezvous.

[redacted] The two UFOs had hardly rejoined when a second object detached and headed straight down toward the ground at high speed. Having regained weapons and communications systems, the aircrew watched the third object, anticipating a large explosion when it struck the ground. However, it landed gently and cast a bright light over a two-three kilometer area. The pilot flew as low over the area as possible, fixing the object's exact location.

[redacted] Upon return to home base, both crewmen had difficulty in

DECLASSIFY ON: 4 Dec 81
by: ACS/I, HQ USAF

33

~~CONFIDENTIAL~~

adjusting their night vision devices for landing. The landing was further complicated by excessive interference on UHF and a further complete loss of all communications when passing through a 150 degree magnetic bearing from the home base. The inertial navigation system simultaneously fluctuated from 30 to 50 degrees. A civil airliner approaching the area also experienced a similar communications failure, but reported no unusual sightings.

[redacted] While on a long final approach, the F-4 crew noted a further UFO. This was described as a cylinder-shaped object (about the size of a T-33 trainer) with bright steady lights on each end and a flasher in the middle. It quickly approached and passed directly over the F-4. In answer to the pilot's query, the control tower reported no other air traffic in the area, although they subsequently obtained a visual sighting of the object when specifically directed where to look.

[redacted] The following day, the F-4 crew was flown by helicopter to the location where they believed the object had landed. This turned out to be a dry lake bed, but nothing unusual was noticed. As the helicopter circled off to the west, however, a very noticeable beeper signal was received, and eventually traced to a nearby house. They immediately landed and asked the inhabitants if anything strange or unusual had occurred the previous night. Yes, they replied, there had been loud noises and a very bright light, like lightning. The helicopter returned to base and arrangements were made to conduct various tests, such as radiation checks, in the vicinity of the house. Unfortunately, the results of such tests have not been reported. ([redacted])

DECLASSIFY ON: 4 Dec 81
ACS/I, HQ USAF

This page intentionally left blank.

Joint Chiefs of Staff Report

Concerning the Sighting of a UFO in Iran on 19 September 1976

This page intentionally left blank.

UNCLASSIFIED

PRIORITY

CT.I MSG654 PAGE 01 257 0813

ACTION: NONE-00

INFO:

ATCZYUW RUEKJCS9712 2660810 MTMS-CCCC--RUFFHQA.

NY CCCCC

2308107 SEP 76

M JCS

NFO RUEHC/SECSTATE WASH DC

UEAIIF/C I A

UEOIAH/NSA WASH DC

UEADWW/WHITE HOUSE WASH DC

UEFHQA/CSAF WASH DC

UENAAA/CNO WASH DC

UEADWD/CSA WASH DC

230630Z SEP 76

M USDAO TEHRAN

O RUEKJCS/DIA WASHDC

INFO RUEKJCS/SECDEF DEPSECDEF WASHDC

RUFRBAA/COMIDEASTFOR

RUDOECA/CINCUSAFE LINDSEY AS GE/INCF

RHFRAAB/CINCUSAFE RAMSTEIN AB GE/INOCN

RUSNAAA/EUDAC VAIHINGEN GER

RUSNAAA/USCINCEUR VAIHINGEN GER/ECJ-2

BT

C O N F I D E N T I A L 1235 SEP76

THIS IS IR 6 846 0139 76

1. (U) IRAN

2. REPORTED UFO SIGHTING (U)

3. (U) NA

4. (U) 19 & 20 SEP 76

5. (U) TEHRAN, IRAN; 20 SEP 76

6. (U) F-6

7. (U) 6 846 0008 (NOTE RO COMMENTS)

8. (U) 6 846 0139.76

9. (U) 22SEP 76

10. (U) NA

11. (U) "INITIATE" IPSP PT-1440

12. (U) USDAO, TEHRAN, IRAN

13. (U) FRANK B. MCKENZIE, COL, USAF, DAT

14. (U) NA

15. (U) THIS REPORT FORWARDS INFORMATION CONCERNING THE SIGHTING OF AN UFO IN IRAN ON 19 SEPTEMBER 1976.

A. AT ABOUT 1230 AM ON 19 SEP 76 THE

RECEIVED FOUR TELEPHONE CALLS FROM CITIZENS LIVING IN THE SHEMIRAN AREA OF TEHRAN SAYING

UNCLASSIFIED

PRIORITY

IORITY

UNCLASSIFIED

TT MSG654 PAGE 02 267 0813

IAT THEY HAD SEEN STRANGE OBJECTS IN THE SKY. SOME REPORTED
KIND OF BIRD-LIKE OBJECT WHILE OTHERS REPORTED A HELICOPTER
TH A LIGHT ON. THERE WERE NO HELICOPTERS AIRBORNE AT THAT
IME. [redacted]

[redacted] AFTER HE TOLD THE CITIZEN IT WAS ONLY
ARS AND HAD TALKED TO MEHRABAD TOWER HE DECIDED TO LOOK FOR
IMSELF. HE NOTICED AN OBJECT IN THE SKY SIMILAR TO A STAR
IGGER AND BRIGHTER. HE DECIDED TO SCRAMBLE AN F-4 FROM
HAHROKHI AFB TO INVESTIGATE.

B. AT 0130 HRS ON THE 19TH THE F-4 TOOK OFF AND PROCEEDED
A POINT ABOUT 40 NM NORTH OF TEHRAN. DUE TO ITS BRILLIANCE
HE OBJECT WAS EASILY VISIBLE FROM 70 MILES AWAY.
S THE F-4 APPROACHED A RANGE OF 25 NM HE LOST ALL INSTRUMENTATION
ND COMMUNICATIONS (UHF AND INTERCOM). HE BROKE OFF THE
NTERCEPT AND HEADED BACK TO SHAHROKHI. WHEN THE F-4 TURNED
AY FROM THE OBJECT AND APPARENTLY WAS NO LONGER A THREAT
O IT THE AIRCRAFT REGAINED ALL INSTRUMENTATION AND COM-
UNICATIONS. AT 0140 HRS A SECOND F-4 WAS LAUNCHED. THE
ACKSEATER ACQUIRED A RADAR LOCK ON AT 27 NM, 12 O'CLOCK
IGH POSITION WITH THE VC (RATE OF CLOSURE) AT 150 NMPH.
S THE RANGE DECREASED TO 25 NM THE OBJECT MOVED AWAY AT A
PEED THAT WAS VISIBLE ON THE RADAR SCOPE AND STAYED AT 25NM.

C. THE SIZE OF THE RADAR RETURN WAS COMPARABLE TO THAT OF
707 TANKER. THE VISUAL SIZE OF THE OBJECT WAS DIFFICULT
O DISCERN BECAUSE OF ITS INTENSE BRILLIANCE. THE
IGHT THAT IT GAVE OFF WAS THAT OF FLASHING STROBE LIGHTS
RRANGED IN A RECTANGULAR PATTERN AND ALTERNATING BLUE, GREEN,
ED AND ORANGE IN COLOR. THE SEQUENCE OF THE LIGHTS WAS SO
AST THAT ALL THE COLORS COULD BE SEEN AT ONCE. THE OBJECT
ND THE PURSUING F-4 CONTINUED ON A COURSE TO THE SOUTH OF
EHRAN WHEN ANOTHER BRIGHTLY LIGHTED OBJECT, ESTIMATED TO BE
NE HALF TO ONE THIRD THE APPARENT SIZE OF THE MOON, CAME
UT OF THE ORIGINAL OBJECT. THIS SECOND OBJECT HEADED STRAIGHT
OWARD THE F-4 AT A VERY FAST RATE OF SPEED. THE PILOT
TTEMPTED TO FIRE AN AIM-9 MISSILE AT THE OBJECT BUT AT THAT
NSTANT HIS WEAPONS CONTROL PANEL WENT OFF AND HE LOST ALL
OMMUNICATIONS (UHF AND INTERPHONE). AT THIS POINT THE PILOT
NITIATED A TURN AND NEGATIVE G DIVE TO GET AWAY. AS HE
URNED THE OBJECT FELL IN TRAIL AT WHAT APPEARED TO BE ABOUT
-4 NM. AS HE CONTINUED IN HIS TURN AWAY FROM THE PRIMARY
BJECT THE SECOND OBJECT WENT TO THE INSIDE OF HIS TURN THEN
ETURNED TO THE PRIMARY OBJECT FOR A PERFECT REJOIN.

D. SHORTLY AFTER THE SECOND OBJECT JOINED UP WITH THE
RIMARY OBJECT ANOTHER OBJECT APPEARED TO COME OUT OF THE

RIORITY

UNCLASSIFIED

PRIORITY UNCLASSIFIED

DCTI: MSG654 PAGE 03 267 0813

OTHER SIDE OF THE PRIMARY OBJECT GOING STRAIGHT DOWN, AT A GREAT RATE OF SPEED. THE F-4 CREW HAD REGAINED COMMUNICATIONS AND THE WEAPONS CONTROL PANEL AND WATCHED THE OBJECT APPROACH THE GROUND ANTICIPATING A LARGE EXPLOSION. THIS OBJECT APPEARED TO COME TO REST GENTLY ON THE EARTH AND CAST A VERY BRIGHT LIGHT OVER AN AREA OF ABOUT 2-3 KILOMETERS.

THE CREW DESCENDED FROM THEIR ALTITUDE OF 26M TO 15M AND CONTINUED TO OBSERVE AND MARK THE OBJECT'S POSITION. THEY HAD SOME DIFFICULTY IN ADJUSTING THEIR NIGHT VISIBILITY FOR LANDING SO AFTER ORBITING MEHRABAD A FEW TIMES THEY WENT OUT FOR A STRAIGHT IN LANDING. THERE WAS A LOT OF INTERFERENCE ON THE UHF AND EACH TIME THEY PASSED THROUGH A MAG. BEARING OF 150 DEGREE FROM EHRABAD THEY LOST THEIR COMMUNICATIONS (UHF AND INTERPHONE) AND THE INS FLUCTUATED FROM 30 DEGREES - 50 DEGREES. THE ONE CIVIL AIRLINER THAT WAS APPROACHING MEHRABAD DURING THIS SAME TIME EXPERIENCED COMMUNICATIONS FAILURE IN THE SAME VICINITY (KILO ZULU) BUT DID NOT REPORT SEEING ANYTHING. WHILE THE F-4 WAS ON A LONG FINAL APPROACH THE CREW NOTICED ANOTHER CYLINDER SHAPED OBJECT (ABOUT THE SIZE OF A T-BIRD AT 10M) WITH BRIGHT STEADY LIGHTS ON EACH END AND A FLASHER IN THE MIDDLE. WHEN QUERIED THE TOWER STATED THERE WAS NO OTHER KNOWN TRAFFIC IN THE AREA. DURING THE TIME THAT THE OBJECT PASSED OVER THE F-4 THE TOWER DID NOT HAVE A VISUAL ON IT BUT PICKED IT UP AFTER THE PILOT TOLD THEM TO LOOK BETWEEN THE MOUNTAINS AND THE REFINERY.

E. DURING DAYLIGHT THE F-4 CREW WAS TAKEN OUT TO THE AREA IN A HELICOPTER WHERE THE OBJECT APPARENTLY HAD LANDED. NOTHING WAS NOTICED AT THE SPOT WHERE THEY THOUGHT THE OBJECT LANDED (A DRY LAKE BED) BUT AS THEY CIRCLED OFF TO THE WEST OF THE AREA THEY PICKED UP A VERY NOTICEABLE BEEPER SIGNAL. AT THE POINT WHERE THE RETURN WAS THE LOUDEST WAS A SMALL HOUSE WITH A GARDEN. THEY LANDED AND ASKED THE PEOPLE WITHIN IF THEY HAD NOTICED ANYTHING STRANGE LAST NIGHT. THE PEOPLE TALKED ABOUT A LOUD NOISE AND A VERY BRIGHT LIGHT LIKE LIGHTENING. THE AIRCRAFT AND AREA WHERE THE OBJECT IS BELIEVED TO HAVE LANDED ARE BEING CHECKED FOR POSSIBLE RADIATION.

MORE INFORMATION WILL BE FORWARDED WHEN IT BECOMES AVAILABLE.

~~XGDS-2-31DEC2000.~~

BT

#9712

PTCCZYUW RUFKJCS9712 2670810 0130-CCCC 2670814

PRIORITY UNCLASSIFIED

This page intentionally left blank.

Routing and Transmittal Slip

from Louis E. Foster, DIA
with a DIA Summary Sheet and
US Defense Attache Report on the
Sighting of a UFO in Iran on
19 September 1976

This page intentionally left blank.

Serial: N9597 E

DIA

ROUTING AND TRANSMITTAL SLIP

Date: 8 Dec 78

TO: (Name, office symbol, room number, building, Agency/Post)	Initials	Date
1. [redacted]		
2.		
3.		
4.		
5.		

	Action		File		Note and Return
	Approval		For Clearance		Per Conversation
	As Requested		For Correction		Prepare Reply
	Circulate		For Your Information		See Me
	Comment		Investigate		Signature
	Coordination		Justify		

REMARKS

Per telecon.

DO NOT use this form as a RECORD of approvals, concurrences, disposals, clearances, and similar actions

FROM: (Name, org. symbol, Agency/Post)	Room No.—Bldg.
Louis E. Foster, RDS-3A, DIA	Phone No. 25766

5041-102
☆ U.S. G.P.O. 1977-241-530/3090

OPTIONAL FORM 41 (Rev. 7-76)
Prescribed by GSA
FPMR (41 CFR) 101-11.206

5. [ ] Releasable Evaluation Not Requested

EVALUATOR'S ORGANIZATION	NAME OF EVALUATOR	DOWNGRADING AND DECLASSIFICATION INSTRUCTIONS
DATE EVALUATED (YEAR / MONTH / DAY)	SIGNATURE OF APPROVING AUTHORITY	Classified by DIA (DB-3)
		SUBJECT TO GENERAL DECLASSIFICATION SCHEDULE OF EXECUTIVE ORDER 11652 AUTOMATICALLY DOWNGRADED AT TWO YEAR INTERVALS DECLASSIFIED ON 31 DECEMBER 1982

CLASSIFICATION CHANGED TO: UNCLASSIFIED

DEFENSE INTELLIGENCE REPORT EVALUATION

INSTRUCTIONS: To be completed by Typewriter or Ball Point Pen

IR NO. 6846 0139 76

COLLECTOR'S PROJECT NUMBER

IR DATE: YEAR 76 MONTH 09 DAY 22

TO:

ICR NUMBER/OTHER REFERENCE

SOURCE NUMBER

DATE REC'D BY DISSEMINATION ACTIVITY: YEAR 76 MONTH 09 DAY 23

SII CODE(S)

TARGET COUNTRY (IES): IR

DATE RECEIVED BY EVALUATOR: YEAR 76 MONTH 10 DAY 1

THRU:

REASON FOR EVALUATION
1. [ ] ICR
2. [ ] COLLECTOR'S REQUEST
3. [X] SELECTED BY COL MGR
4. [ ] SPECIAL REQUEST
5. [ ] ANALYST INITIATIVE

IR ORIGINATOR: DIA IR

A. IR RESPONDED TO
1. [X] DIRM Part Three
2. [ ] CIR
3. [ ] ICR—If checked, ICR was satisfied:
a. [ ] Completely
b. [ ] Partially (If b or c checked, include Guidance in Remarks Section)
c. [ ] Not at all
4. [ ] Other

B. RELIABILITY OF INFORMATION
1. [X] Confirmed by other sources
2. [ ] Substantially true
3. [ ] Cannot be judged
4. [ ] Doubtful
5. [ ] False

C. VALUE OF INFORMATION
1. [X] High (Unique, Timely, and of Major Significance)
2. [ ] Moderate (Contributory and Useful)
3. [ ] Low (Marginal)
4. [ ] None (Of no use)
5. [ ] Cannot be judged (Analyst has no basis for value judgement)

D. USABILITY OF INFORMATION
1. [ ] Used or planned for use in product:
a. [ ] Basic Intelligence
b. [X] Current Intelligence
c. [ ] Estimative Intelligence
d. [ ] Other

2. [ ] Incorporated in Data Base
a. [X] Potentially Useful
b. [ ] Background/Confirmatory
3. [ ] Stimulus for Intelligence Guidance or Requirement

4. [ ] Not used (All responses require explanation in Remarks section)
a. [ ] Unreliable
b. [ ] Too Fragmentary
c. [ ] Duplication
d. [ ] Not Pertinent to Needs

NAME OF PRODUCT:

REMARKS (Number and Classify each Paragraph)

1. An outstanding report. This case is a classic which meets all the criteria necessary for a valid study of a UFO phenomenon:

a) The object was seen by multiple witnesses from different locations (i.e. Shahrokhi, Mehrabad, and the dry lake bed) and viewpoints (both airborne and from the ground).

b) The credibility of many of the witnesses was high (an air force general, qualified aircrews, and experienced tower operators).

c) Visual sightings were confirmed by radar.

d) Similar electromagnetic effects (EME) were reported by three separate aircraft.

e) There were physiological effects on some crew members (i.e. loss of night vision due to the brightness of the object).

f) An inordinate amount of maneuver-ability was displayed by the UFO's.

EVALUATOR'S RELEASE and DISCLOSURE GUIDANCE
1. [ ] Paragraph(s) ______ above are releasable to the government of ______
2. [ ] Evaluation is Not Releasable
3. [ ] Releasable Evaluation Not Requested

EVALUATOR'S ORGANIZATION

NAME OF EVALUATOR

DATE EVALUATED: YEAR 76 MONTH 10 DAY 12

SIGNATURE OF APPROVING AUTHORITY: [signature]

DOWNGRADING AND DECLASSIFICATION INSTRUCTIONS
Classified by DIA-DB-3E
SUBJECT TO GENERAL DECLASSIFICATION SCHEDULE OF EXECUTIVE ORDER 11652 AUTOMATICALLY DOWNGRADED AT TWO YEAR INTERVALS DECLASSIFIED ON 31 DECEMBER 1982

DD FORM 1131

EDITION OF ... IS OBSOLETE.

CLASSIFICATION

CLASSIFICATION CHANGED TO: UNCLASSIFIED

DEPARTMENT OF DEFENSE
JOINT CHIEFS OF STAFF
MESSAGE CENTER

VZCZCMAY80&ILN118 [redacted]
MULT 2843
ACTION
DIA:
DISTR
CJCS(01) DJS(03) J3(14) J5(02) NMCC SECDEF(07) DPSECDEF
NMIC SECDEF: ASD:ISA(10) :DIA(15)
* SECSTATE WASH DC
* C I A
* NSA WASH DC
* WHITE HOUSE WASH DC
CMC
* CSAF WASH DC
* CNO WASH DC
* CSA WASH DC
FILE(1)
(053)

TRANSIT/2306302/230810Z/001:40TOR2670804
DE RUQMHRA #9575 2670615
ZNY CCCCC
P 230630Z SEP 76
FM USDAO TEHRAN
TO RUEKJCS/DIA WASHDC
INFO RUEKJCS/SECDEF DEPSECDEF WASHDC
RUFRBAA/COMIDEASTFOR
RUDOECA/CINCUSAFE LINDSEY AS GE/INCF
RMFRAAB/CINCUSAFE RAMSTEIN AB GE/INOCN
RUSNAAA/EUDAC VAIHINGEN GER
RUSNAAA/USCINCEUR VAIHINGEN GER/ECJ-2
BT
[redacted] 1235 SEP76
THIS IS IR 6 846 0139 76
1. (U) IRAN
2. REPORTED UFO SIGHTING (U)
3. (U) NA
4. (U) 19 & 20 SEP 76
5. (U) TEHRAN, IRAN 20 SEP 76
6. (U) F-6
7. (U) 6 846 0000 (NOTE RO COMMENTS)
8. (U) 6 846 0139 76
9. (U) 22SEP 76
10. (U) NA
11. (U) "INITIATE" IPSP PT-1440

A

379/80-IR

098

PAGE 1 [redacted] 00110101 53

REC'D 22 SEP 76

(45)

PAGE 2 [redacted] 12843
12. (U) USDAO, TEHRAN, IRAN
13. (U) FRANK B. MCKENZIE, COL, USAF, DATT
14. (U) NA
15. [redacted] THIS REPORT FORWARDS INFORMATION CONCERNING THE SIGHTING OF AN UFO IN IRAN ON 19 SEPTEMBER 1976.

A. AT ABOUT 1230 AM ON 19 SEP 76 THE IMPERIDAL IRANIAN AIR FORCE (IIAF) COMMAND POST RECEIVED FOUR TELEPHONE CALLS FROM CITIZENS LIVING IN THE SHEMIRAN AREA OF TEHRAN SAYING THAT THEY HAD SEEN STRANGE OBJECTS IN THE SKY. SOME REPORTED A KIND OF BIRD-LIKE OBJECT WHILE OTHERS REPORTED A HELICOPTER WITH A LIGHT ON. THERE WERE NO HELICOPTERS AIRBORNE AT THAT TIME. THE COMMAND POST CALLED BG YOUSEFI, ASSISTANT DEPUTY COMMANDER OF OPERATIONS. AFTER HE TOLD THE CITIZEN IT WAS ONLY STARS AND HAD TALKED TO MEHRABAD TOWER HE DECIDED TO LOOK FOR HIMSELF. HE NOTICED AN OBJECT IN THE SKY SIMILAR TO A STAR BIGGER AND BRIGHTER. HE DECIDED TO SCRAMBLE AN F-4 FROM SHAHROKHI AFB TO INVESTIGATE.

B. AT 0130 HRS ON THE 19TH THE F-4 TOOK OFF AND PROCEEDED TO A POINT ABOUT 40 NM NORTH OF TEHRAN. DUE TO ITS BRILLIANCE THE OBJECT WAS EASILY VISIBLE FROM 70 MILES AWAY. AS THE F-4 APPROACHED A RANGE OF 25 NM HE LOST ALL INSTRUMENTATION AND COMMUNICATIONS (UHF AND INTERCOM). HE BROKE OFF THE INTERCEPT AND HEADED BACK TO SHAHROKHI. WHEN THE F-4 TURNED AWAY FROM THE OBJECT AND APPARENTLY WAS NO LONGER A THREAT TO IT THE AIRCRAFT REGAINED ALL INSTRUMENTATION AND COMMUNICATIONS. AT 0140 HRS A SECOND F-4 WAS LAUNCHED. THE BACKSEATER ACQUIRED A RADAR LOCK ON AT 27 NM, 12 O'CLOCK HIGH POSITION WITH THE VC (RATE OF CLOSURE) AT 150 NMPH. AS THE RANGE DECREASED TO 25 NM THE OBJECT MOVED AWAY AT A SPEED THAT WAS VISIBLE ON THE RADAR SCOPE AND STAYED AT 25NM.

C. THE SIZE OF THE RADAR RETURN WAS COMPARABLE TO THAT OF A 707 TANKER. THE VISUAL SIZE OF THE OBJECT WAS DIFFICULT TO DISCERN BECAUSE OF ITS INTENSE BRILLIANCE. THE LIGHT THAT IT GAVE OFF WAS THAT OF FLASHING STROBE LIGHTS ARRANGED IN A RECTANGULAR PATTERN AND ALTERNATING BLUE, GREEN, RED AND ORANGE IN COLOR. THE SEQUENCE OF THE LIGHTS WAS SO FAST THAT ALL THE COLORS COULD BE SEEN AT ONCE. THE OBJECT AND THE PURSUING F-4 CONTINUED ON A COURSE TO THE SOUTH OF TEHRAN WHEN ANOTHER BRIGHTLY LIGHTED OBJECT, ESTIMATED TO BE ONE HALF TO ONE THIRD THE APPARENT SIZE OF THE MOON, CAME OUT OF THE ORIGINAL OBJECT. THIS SECOND OBJECT HEADED STRAIGHT TOWARD THE F-4 AT A VERY FAST RATE OF SPEED. THE PILOT ATTEMPTED TO FIRE AN AIM-9 MISSILE AT THE OBJECT BUT AT THAT

INSTANT HIS WEAPONS CONTROL PANEL WENT OFF AND HE LOST ALL COMMUNICATIONS (UHF AND INTERPHONE). AT THIS POINT THE PILOT INITIATED A TURN AND NEGATIVE G DIVE TO GET AWAY. AS HE TURNED THE OBJEAZ FELL IN TRAIL AT WHAT APPEARED TO BE ABOUT 3-4 NM. AS HE CONTINUED IN HIS TURN AWAY FROM THE PRIMARY OBJECT THE SECOND OBJECT WENT TO THE INSIDE OF HIS TURN THEN RETURNED TO THE PRIMARY OBJECT FOR A PERFECT REJOIN.

D. SHORTLY AFTER THE SECOND OBJECT JOINED UP WITH THE PRIMARY OBJECT ANOTHER OBJECT APPEARED TO COME OUT OF THE OTHER SIDE OF THE PRIMARY OBJECT GOING STRAIGHT DOWN, AT A GREAT RATE OF SPEED. THE F-4 CREW HAD REGAINED COMMUNICATIONS AND THE WEAPONS CONTROL PANEL AND WATCHED THE OBJECT APPROACH THE GROUND ANTICIPATING A LARGE EXPLOSION. THIS OBJECT APPEARED TO COME TO REST GENTLY ON THE EARTH AND CAST A VERY BRIGHT LIGHT OVER AN AREA OF ABOUT 2-3 KILOMETERS.

THE CREW DESCENDED FROM THEIR ALTITUDE OF 26M TO 15M AND CONTINUED TO OBSERVE AND MARK THE OBJECT'S POSITION. THEY HAD SOME DIFFICULTY IN ADJUSTING THEIR NIGHT VISIBILITY FOR LANDING SO AFTER ORBITING MEHRABAD A FEW TIMES THEY WENT OUT FOR A STRAIGHT IN LANDING. THERE WAS A LOT OF INTERFERENCE ON THE UHF AND EACH TIME THEY PASSED THROUGH A MAG. BEARING OF 150 DEGREE FROM EHRABAD THEY LOST THEIR COMMUNICATIONS (UHF AND INTERPHONE) AND THE INS FLUCTUATED FROM 30 DEGREES - 50 DEGREES. THE ONE CIVIL AIRLINER THAT WAS APPROACHING MEHRABAD DURING THIS SAME TIME EXPERIENCED COMMUNICATIONS FAILURE IN THE SAME VICINITY (KILO ZULU) BUT DID NOT REPORT SEEING ANYTHING.

WHILE THE F-4 WAS ON A LONG FINAL APPROACH THE CREW NOTICED ANOTHER CYLINDER SHAPED OBJECT (ABOUT THE SIZE OF A T-BIRD AT 10M) WITH BRIGHT STEADY LIGHTS ON EACH END AND A FLASHER IN THE MIDDLE. WHEN QUERIED THE TOWER STATED THERE WAS NO OTHER KNOWN TRAFFIC IN THE AREA. DURING THE TIME THAT THE OBJECT PASSED OVER THE F-4 THE TOWER DID NOT HAVE A VISUAL ON IT BUT PICKED IT UP AFTER THE PILOT TOLD THEM TO LOOK BETWEEN THE MOUNTAINS AND THE REFINERY.

E. DURING DAYLIGHT THE F-4 CREW WAS TAKEN OUT TO THE AREA IN A HELICOPTER WHERE THE OBJECT APPARENTLY HAD LANDED. NOTHING WAS NOTICED AT THE SPOT WHERE THEY THOUGHT THE OBJECT LANDED (A DRY LAKE BED) BUT AS THEY CIRCLED OFF TO THE WEST OF THE AREA THEY PICKED UP A VERY NOTICEABLE BEEPER SIGNAL. AT THE POINT WHERE THE RETURN WAS THE LOUDEST WAS A SMALL HOUSE WITH A GARDEN. THEY LANDED AND ASKED THE PEOPLE WITHIN IF THEY HAD NOTICED ANYTHING STRANGE LAST NIGHT. THE PEOPLE TALKED ABOUT A LOUD NOISE AND A VERY BRIGHT LIGHT

DEPARTMENT OF DEFENSE
JOINT CHIEFS OF STAFF
MESSAGE CENTER

PAGE 4 = = - [redacted] 12043
LIKE LIGHTENING. THE AIRCRAFT AND AREA WHERE THE OBJECT IS
BELWEVED TO HAVE LANDED ARE BEING CHECKED FOR POSSIBLE RADIATION.
RO COMMENTS: [redacted] ACTUAL INFORMATION CONTAINED IN THIS REPORT
WAS OBTAINED FROM SOURCE IN CONVERSATION WITH A SUB-SOURCE, AND
IIAF PILOT OF ONE OF THE F-4S. MORE INFORMATION WILL BE
FORWARDED WHEN IT BECOMES AVAILABLE.
[redacted]
BT
#9575
ANNOTES
JEP 117

23 SEP 1976 13 11
DIA-DS-3C
OX5-5865

PAGE 4 [redacted] 00110101

NNNN
2308107

Exploitation Report

A Fragment, Metal
Recovered in the Republic of the Congo

This page intentionally left blank.

CON[illegible]

EXPLOITATION REPORT

ON FRAGMENT, METAL, RECOVERED IN THE REPUBLIC OF THE CONGO, [illegible] ORIGIN BELIEVED TO BE AN UNIDENTIFIED FLYING OBJECT [illegible] (COUNTRY UNIDENTIFIED) (U[illegible]

[illegible]-25500

SECTION I. (C) Purpose (U[illegible]

1. (C) The purpose of this report is to present the results of the exploitation of a metalic fragment recov[illegible] near the town of Karakana in the Republic of the Congo. [illegible] recovery was the result of a ground-level search which was co[illegible] after an unidentified flying object exploded and fell to earth in the area. The sighting and recovery took place sometime between 10 and 15 October 1963. Other than a reported east-to-west direction of flight for the UFO, specific observation and recovery details are lacking.

SECTION II. (C) Description [illegible]

2. (C) Details concerning the exact lo[illegible] and characteristics of impact are unknown. However, the appearance of the fragment indicated exposure to high temperatures prior [illegible] (earth). [illegible] impact of the specimen had little or no effect on its final condition or appearance. The fragment weighed [illegible] had an indefinite density of approximately iron and measured 2.25 x 1.75 x 1.0 inches. The top and side views of the specimen were rounded and appeared to have been shaped by heating and melting. This is illustrated in Figures 1 and 2. The [illegible] shaped groove, visible in Figure 1, is the outline of an insert of shade that differs [illegible] from the rest of the specimen. Figure 3 shows the side opposite the end shown in Figure 1.

GROUP [illegible]
[illegible] YEAR INTERVALS
[illegible] DECLASSIFIED

CON[illegible]

REGRADED UNCLASSIFIED
ON 1 6 FEB 1992
BY CDR USAINSCOM FOI/PO
AUTH Para 1-603 DOD 5200.1R

POOR ORIGINAL

CONFIDENTIAL

and was composed of six machined or formed "T" shaped fins extending along the major axis of the fragment.

SECTION III. (C) Conclusions

3. (C) The fragment was originally pr[illegible] an electrical component and could be identified as a mot[illegible] ature, generator armature, or associated electrical regul[illegible] device.

4. (C) The fragment was constructed [illegible].010-inch thick silicon steel laminate stacked on a central mild steel core or shaft.

5. (C) Materials, processes, dimensions, etc., as such, prevent determination of exact origin (country).

6. (C) Surface appearance and microstructure of the specimen indicates exposure to temperatures in excess of 2500° F.

SECTION IV. (C) Exploitatio[illegible] Data (U)

7. (C) The recovered spe[illegible] wei[illegible].41 grams and ha[illegible] a density closely approximati[illegible] [illegible] flowed metal or heavy oxide was [illegible] temperatures in excess of 2500° F. While there are no indications of impact, the flowed metal, as shown in Figures 4 and 7, would substantiate the conclusion that the item was moving at a high velocity when it was hot.

8. (C) Fabrication of the item was accomplished utilizing more or less standard procedures for fabricating electric motor armatures. Armature laminates were stamped (punched) from approximately .012-inch sheet steel, copperplated, and assembled on a mild steel shaft approximately .405 inches in diameter. Following assembly, the laminates were joined by solid-state or diffusion-bonding of the copper-

REGRADED UNCLASSIFIED
ON 1 0 FEB 1999
BY CDR USAINSCOM FOI/PO
AUTH Para 1-603 DOD 5200.1R

CONFIDENTIAL

POOR ORIGINAL

CON[illegible]

placing. This can be accomplished by tightly compacting the laminate assembly and heating in a furnace. Temperature required for bonding of the copper depends upon the degree compaction or pressure; the higher pressures requiring proportionately lower temperatures.

9. (C) A cross-section (transverse to [illegible] length of the specimen) is shown in Figure 5. The light-colored, [illegible] lines are the edges of individual laminates, caused by cutting at [illegible] angle to, instead of parallel to, the laminates. [illegible] that [illegible] fins or petals are "T" shaped. This shape is used to help [illegible] winding wire in place and is found on high RPM motors. The melted condition of some of the "T's" is indicative of the high heating conditions experienced. The outer surface of the armature shaft is serrated to prevent axial slippage of the laminates.

10. (C) The lamination or stacking of [illegible] individual laminates is clearly illustrated in Figure 6. The [illegible] laminates on the fin at the top of the photograph is due to the melting and [illegible] of the copperplating during the high temper[illegible] of the specimen. Some of the copper has been [illegible] the fin at the bottom of the photograph. A [illegible] of this area is shown in Figures 7 and 8.

11. (C) Another result of intense heating was the enlarged grain size of the steel laminates' microstructure. The [illegible] of the laminates shown in Figures 9 and 10 illustrates grain sizes that are comparable to Category 2 of ASTM [illegible]. [illegible] large grain size [illegible]

REGRADED UNCLASSIFIED
ON 10 FEB 1992
BY CDR USAINSCOM F01/PO
AUTH Para 1-603 DOD 5200.1R

CO[illegible]

CONFIDENTIAL

intense heat and then cooled at a comparatively slow rate.

12. (C) The light material between the laminations in Figure 9 is plated copper that melted and flowed between the laminations when the entire specimen was hot. A photomicrograph of this is shown in Figure 11.

13. (C) Analysis of the [illegible] disclosed the following:

Element	Percent Present (Weight)
Carbon	[illegible]
Manganese	[illegible]
Silicon	[illegible]
Nickel	less than 0.10
Chromium	0.37
Molybdenum	less than 0.01

14. (C) Chemical composition of the steel laminations was as follows:

Element	Percent Present (Weight)
Manganese	25
Silicon	[illegible]
Nickel	[illegible]
Chromium	[illegible]
Molybdenum	less than 0.01

CONFIDENTIAL

Figure 1 [illegible] Top View [illegible] 265 [illegible]

Figure 2 [illegible] Side View of [illegible]

UNCLASSIFIED

Figure 3.(U)Opposite View of Fig. 1 (U)

Figure 4.(U) Flawed [illegible] Head, [illegible]
Magnification [illegible] 2X (U)

6

UNCLASSIFIED

Figure 3 (U) Cross Section (Transverse to the Length of the Specimen) Magnification 21 (U)

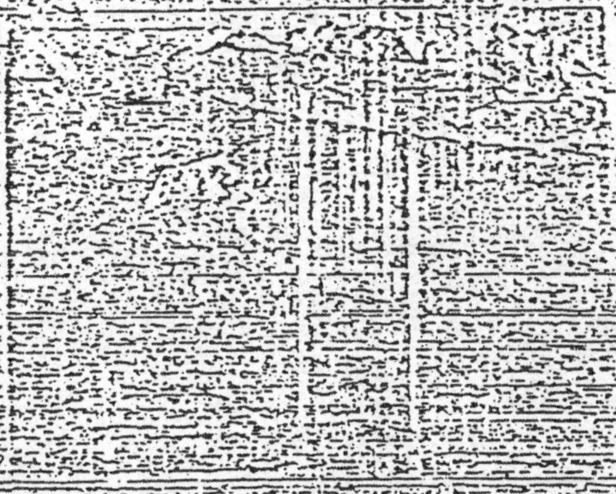

UNCLASSIFIED

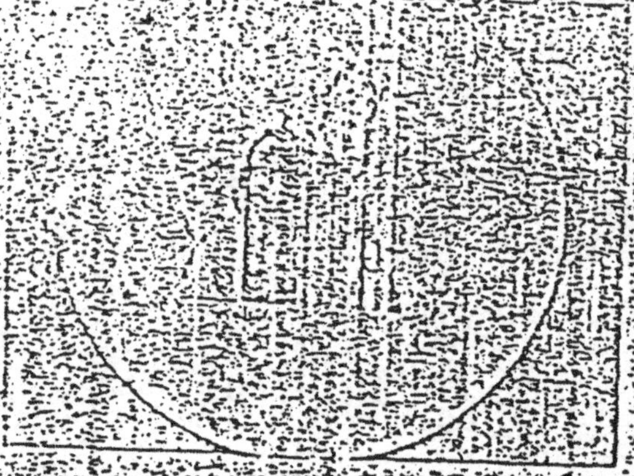

Figure 7. (U) Macrograph Cross-Section of Laminated Area Magnification 1.75X (U)

Figure 8 (U) Micrograph Cross-Section of Laminated Area Magnification [illegible] (U)

8

UNCLASSIFIED

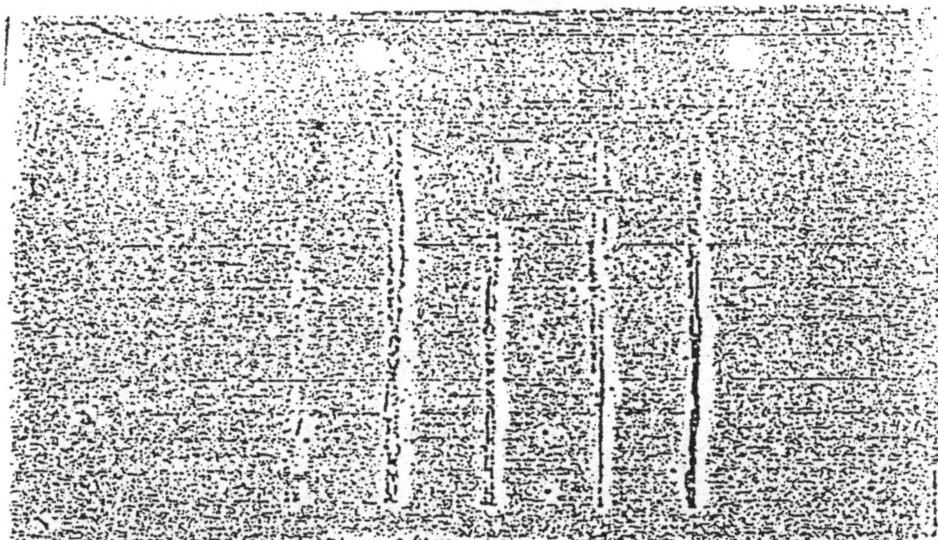

Figure 9 (a) [illegible]
[illegible]
Magnification 63
(3% Nital Etch[illegible]

Figure 1[illegible]

UNCL[illegible]

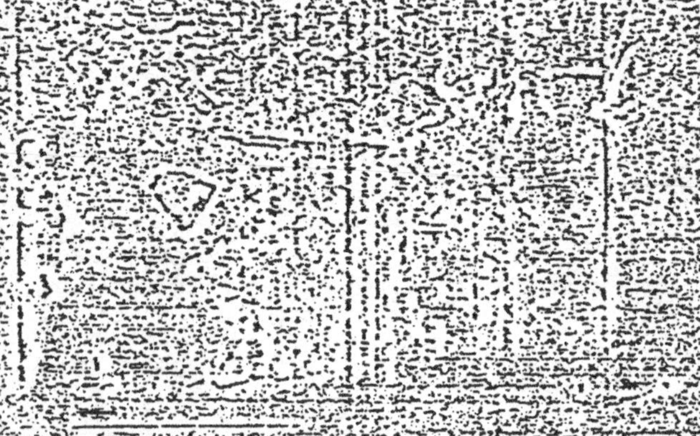

Figure 11.(U) Photomicrograph of Light [illegible] Material Between the Laminations
Magnification 790X
(Mixed Acids Etchant)(U)

13

U

ORIGINAL

Navy Report

Country: Cuba
Subject: UFO

This page intentionally left blank.

UNCLASSIFIED

I58004-35D

Country: CUBA
Subject: UFO (UNIDENTIFIED FLYING OBJECT) (U)
D.I.: 10 AUGUST 1968
Pl & Date Acq: U.S. NAVAL BASE GTMO
10 AUGUST 1968
Eval: F-6
Source: USS BARRY (DD 933)
USS LASALLE (LPD 3)

Info Spec: EES/bk
Distr: EXP/EMT(J)

MF Reel 2423 Frame
Rpt: 5-366-0326-68
D.R.: 13 AUGUST 1968
No. Pages: 1
Ref: A. CINCLANTFLT 071718Z AUG 68
B. NAS GTMO 101723Z AUG 68
Originator: COMNAVBASE GTMO
Prep by: LCDR J. GIACOMUZZI
Appr Auth: R. M. HARRIS
Chief of Staff

Entire Report

SUMMARY:

THIS REPORT DESCRIBES THE SIGHTING OF A UFO. THE POSSIBILITY EXISTS THAT IT MAY BE A DECAYING SATELLITE AS INDICATED IN REFERENCE A.

REPORT:

1. REFERENCE A REPORTED A SATELLITE OF COSMOS 236 DECAYING AND INDICATED VARIOUS TRACK POSITIONS FOR POSSIBLE SIGHTINGS. REFERENCE B. REPORTED A PROBABLE SIGHTING.

2. THE SUBJECT UFO WAS SIGHTED AT 110028Z BY NUMEROUS PERSONNEL ABOARD SHIP IN THE GTMO AREA AND BY OTHERS ABOARD THE NAVAL BASE. THE OBJECT WAS SIGHTED TO THE NORTH OF NAVBASE GTMO APPROXIMATELY 60 DEGREES ABOVE THE HORIZON, AND ON A COURSE OF 130 DEGREES TRUE. THE OBJECT WAS GLOWING BRIGHTLY AND EMITTED WHAT APPEARED TO BE A SPIRAL TRAIL OF SMOKE OR VAPOR.

1/2

UNCLASSIFIED

I58004-35D
CUBA

Rpt: 5-366-0326-68

3. CREWMEN ABOARD THE USS BARRY AND LASALLE MANAGED TO OBTAIN KODACHROME PHOTOGRAPHS OF THE SUBJECT OBJECT. IN THIS RESPECT, ATTACHMENTS (1) AND (2) ARE HEREBY FORWARDED, FOR DEVELOPING AND INFORMATION.

ATTACH:
(1) ONE ROLL OF 35MM KODACHROME II COLOR FILM (36 EXP.) [Not received]
(2) ONE CARTRIDGE OF KODACHROME II COLOR FILM (8MM) [Not received]
[Received 10 color photos]
REQUEST PHOTOS DIRECTLY FROM DIAAP-10X

2/2

This page intentionally left blank.

Department of State AIRGRAM

1. Subject:
   Flying Saucers Are a Myth
2. Subject:
   (DOS Report) Tunisian Firefall
3. Subject:
   Celestial Body Observed over Antarctica
4. Subject:
   Political Report for August 1965
5. Subject:
   Unidentified Flying Objects
6. Subject:
   Sighting of Object Possibly Originating from Space Vehicle
7. Subject:
   Unidentified Flying Objects
8. Subject:
   Unidentified Flying Objects Reported Over Angola
9. Subject:
   Report of Unidentified Object Which Fell Near Municipio of General Teran, N.L.

This page intentionally left blank.

DEPARTMENT OF STATE
FPC/RD

(N1)

DEPARTMENT OF STATE MAY 19 8 15 AM '95

AIRGRAM

SP 16

FOR RM USE ONLY

20 (ENCL REPRODUCED IN FBR)

RM/R	REP	AF
1		
ARA	EUR	FE
	5	
NEA	CU	INR
		5
E	P	IO
4	2	5
L	FBO	AID
SY	S/P	G/PM
1	1	1
AGR	COM	FRB
	10	
INT	LAB	TAR
7		
TR	SMI	AIR
	1	5
ARMY	CIA	NAVY
3	20	5
OSD	USIA	NSA
34	10	3
AEC	NIC	NSC
8	1	6
NASA	NSF	OST
4	4	1

A-1221 — UNCLASSIFIED

NO. — HANDLING INDICATOR

TO : DEPARTMENT OF STATE

S131

~~P22343~~
P22121

RECEIVED

MAR 24 10 59 AM 1968

ANALYSIS BRANCH

FROM : AmEmbassy MOSCOW — DATE: March 22, 1968

SUBJECT : Flying Saucers Are a Myth (or maybe a Mythis)

REF : Moscow's A-1095, Feb. 20, 1968

A long article in Moskovskiy Komsomolets of February 16, 1968, by Vil LYUSTIBERG, Science Editor of the Novosti Press Agency, debunks flying saucers completely. Flying saucers, says Lyustiberg, appear to those who believe in them, but persistently stay away from air observation posts, meteorologists and astronomers. Lyustiberg, in fact, is quite emphatic throughout his article that unidentified flying objects (UFO) do not exist. He makes no attempt to square this belief with previous published Soviet articles, including that rather spectacular article primarily for U.S. consumption in Soviet Life (see reference).

THOMPSON

1968 MAR 24 AM 11 36
COPYFLO-PBR

Enclosure: "Flying Saucers? They're a Myth"

UNCLASSIFIED

FOR DEPT. USE
[X] In [ ] Out

FORM 4-62 DS-323

Drafted by: SCI:CASquire:hz 3/21/68

Contents and Classification Approved by: First Secretary of Embassy: CASquire

Clearances: MIN:ECSwank

Enclosure to
Moscow A-1221
6 scl

Tuesday, March 12, 1968

"FLYING SAUCERS"? THEY'RE A MYTH!

Villen Lyustiberg, APN Science Commentator

In the last 20 years publication of literature on "unidentified flying objects" has advanced to one of the first places in the world. This problem is dealt with by numerous associations and clubs.

Are these "flying saucers" just imagination or reality? Thousands of recorded statements by eye-witnesses seem to show that "there must be something in it". But let us try to look at all these materials from a different stand. So, as the convinced supporters of the "saucer" theory say almost every time, you can draw your own conclusions, while we give you.

"Facts Only"

Kenneth Arnold, an American pilot, is the "godfather" of flying saucers. It was he who noticed in 1947 nine shining discs flying in formation at the speed of about 3,000 km an hour.

The next encounter with saucers had a tragic outcome. In January 1948, an immense flying saucer appeared over the Nox US Air Base. Pursuing it Captain Thomas Manthell perished. The last thing he reported from an altitude of 9,000 metres was that he saw an object and was going to approach it.

This story found extensive response. Many considered it to be a stern warning of unknown envoys from other planets to leave them alone. There was a great deal of a talk about the mysterious death of the pilot. It was received as undoubted proof of the existence of mighty forces still unknown to us.

But the results of a thorough investigation by US authorities received much less publicity. The mysterious "saucer" proved to be nothing but a thin-walled plastic balloon made by US Naval Forces under the secret "Skyhawk Operation" project.

Such balloons could rise to an altitude of 30,000 metres. Meanwhile, Thomas Manthell pursuing it, forgot that he had no oxygen apparatus on board his craft.

We know of stories about encounters and even hand-to-hand fights with pilots of craft landing from other planets. They were shot at, but without success.

A saucer cut down a tree on the Amazon River shore and disappeared in the turbid stream. It was never recovered. Another saucer glided over the earth like an injured bird and almost crashed before the eyes of witnesses. But "having spit out" several pieces of metal, it levelled out its course and flew off. Delivered to the police, the metal proved to be ordinary tin.

PLANT →

An abandoned silvery disc was found in the deep rock-coal seams in Norwegian coal mines on Spitzbergen. It was pierced and marked by micrometeor impacts and bore all traces of having performed a long space voyage. It was sent for analysis to the Pentagon and disappeared there.

Nothing but a saucer put out of commission a high-voltage power transmission line in 1965 and thus plunged several large American cities into darkness for six hours.

But the most thrilling masterpiece of this sort was probably the "Interview with a Man from Venus" published at the close of 1967 by the West German Stern magazine. This materialized blue-eyed "superman", a version of Nietzsche's "blond beast", proved to be a secret service agent of the Pentagon. He was 190 cm tall, spoke excellent English (it was English indeed!) and could breathe freely in our atmosphere without any devices. His modest fibre suitcase contained an unusual silvery suit, flexible like silk and so hard that a diamond drill broke off it.

The man from Venus said that he could walk freely in such a suit over the surface of the hottest stars, ignoring all powerful gravitational fields.

Maybe these "facts" will do?

How Can They Be Studied?

It is easier to ask this question, than to answer it. These flying saucers -- they are like Our Lady: they appear to those who believe in them. And they persistently fail to show themselves to air observation posts, meteorologists and astronomers, i.e. precisely to those who can give us accurate information and trustworthy descriptions of a flying object.

Experiments that do not repeat themselves, or the chance appearance of a phenomenon always either handicap the possibility of investigation considerably or exclude it altogether.

Nowhere in the world is there EVEN ONE trustworthy stereoscopic photograph making it possible to trace from two points simultaneously the outlines of a solid flying object clearly. All photographs that exist have been made with single-lens cameras and always leave a margin for doubt.

A series of such photographs amazingly resembles a straw hat with a black silk ribbon, thrown into the air. Others undoubtedly prove to be internegatives obtained from two combined slides, for instance, that of a landscape and that of a strangely designed craft drawn on white paper. Modern photographing techniques allow photographs to be produced which no expert will ever prove to be counterfeit. They are real. But they do not show strangers from space; they show quite ordinary earthly objects unexpectedly foreshortened — pots, pans, plastic toys, and the like.

Are there photographs which evoke no doubts whatsoever? Yes, there are. They always show shining objects of a circular or oval shape with vague outlines. Most likely these photographs show ball lightning. Incidentally, the nature of the latter has not been fully elucidated yet either, due to the vagueness of the place and time of its formation. However, no one ever thought of ascribing a cosmic origin to it.

Even with a superficial analysis, mysterious flares on clouds proved to be reflections of electric welding, warning lights of airplanes in the area of airports, or distant summer lightning. And the November 1967 "saucer" over Sofia proved to be a high-altitude NATO reconnaissance balloon.

Among the most serious works devoted to an investigation of the problem, there are two which ought to be mentioned. One of them is the book "Flying Saucers" by Donald Menzel who explains almost all cases of their appearance by disturbances in the Earth's or Sun's atmosphere. Our planet travels in the upper layers of the Sun's atmosphere where clots of high-temperature plasma are moving freely.

The other is a book by Frank Edwards. It is a collection of statements, notes and records of eye-witness testimony classified into several sections. True, Edwards often qualifies his source as a witness who did not wish to have his name mentioned, but who may be fully trusted, or that the trustworthiness of the witness is confirmed by numerous of his countrymen, but he cannot mention his name so far for a number of reasons. That's not very convincing, is it?

For Whom Is It Not a Myth?

They are those for whom science is a business. Taking advantage of the lively interest of people for everything that is strange and unusual, numerous lecturers in the West appear before audiences with reports and stories, invite eye-witnesses to such lectures, and demonstrate photographs and slides. Most of these lecturers are nothing but ordinary quacks. True, there are people who are sincerely convinced in the truth of what they are talking about. They strive to draw the interest of the public to certain phenomena but, we are sorry to say, there is more harm than use in that.

The Americans Lesly and Adamski are the most outright frauds among them. Using their "good friendly relations" with strangers from other planets, the enterprising businessmen "visited" Mars, the Moon and Venus, learned from our cosmic neighbours to treat by simple and accessible methods such diseases as cancer, glaucoma, hypertension and others which afflict the human race. Their lectures, motion-pictures, books and medical practice have brought them many hundred thousand dollars out of the pockets of trusting listeners and patients already.

We may trace a clearly defined regularity in the appearance of large numbers of flying saucers. And, strange as it might seem, this regularity is closely connected with earthly events.

The first "cycle" of four years began in 1947. The number of saucers always grows sharply on the eve of presidential elections in the USA. This is difficult to explain. Maybe people on other planets lay bets as to who will win in the next elections -- the republicans or the democrats. Perhaps, these saucers appear in order to divert the voters' thoughts from the again non-fulfilled presidential programme and promises to make the country a "great" and "prospering" society.

There is another cycle. True, it is not determined by time. Saucers flooded the earth's atmosphere in 1951, in the months when the American troops were waging a ruthless, annihilating war against the Korean People's Democratic Republic.

The next invasion of saucers was in 1965-1966, when 35 American advisers in South Vietnam were quickly replaced by half-a-million-strong army equipped with the latest arms and intending to rout the National-Liberation Front within two weeks and when the President of the United States of America, without declaring war and in violation the constitution of his own country, gave the order that the peaceful population of North Vietnam be subjected to brutal bombings.

In short, when newspaper circulation drops, when readers get tired of economics and politics and when they are to be diverted from "irrelevant" questions, the Western businessmen resort to three reliable, "always fresh" sensations: Flying Saucers, the Sea Serpent (sometimes it is substituted by the Loch Ness Lake monster), and the Snow Man.

It is much nicer to read about mysterious craft from Venus than to think of the future, of the wage freeze, of growing prices or unemployment. Statesmen in imperialist countries resort to this "information" quite deliberately. For them the flying saucers are not a myth, but a well-camouflaged means for misinforming the people. And nothing but that.

(Moskovsky Komsomolets, February 16. In full.)

THE END

UNCLASSIFIED

CLASSIFIED CLASSIFIED CLASSIFIED CLASSIFIED

N2

[redacted]

S131

UNCLASSIFIED CA0338

PAGE 01 TUNIS 03293 141307Z

15

ACTION SCI 05

INFO OCT 01, AF 12, SMI 01, ACDA 16, CIAE 00, DODE 00, JPH 04, INR 07, L 03,
SAE 00, NASA 04, NSC 10, OST 01, RSC 01, P 04, USIA 12, AEC 11, NSA 01,
/093 W

--------------------- 002517

141265Z JUL 69

AMEMBASSY TUNIS

SECSTATE WASHDC 9587

4

TUNIS 3293

SMITHSONIAN OBSERVATORY CAMBRIDGE

N3

FOR OCI

SUBJ: TUNISIAN FIREFALL

REF: TUNIS 3222

1. PHENOMINA ALMOST IDENTICAL THAT OF REFTEL SEEN BY MANY OBSERVERS TUNIS EVENING JULY 13 INCLUDING EMBASSY OFFICIALS. AGAIN CONSISTED OF TWO PARTS ONE GREENISH-BLUE CIRCULAR CLOUD ABOUT SIZE OF FULL MOON APPEARING A FEW DEGREES TO WEST OF POLAR STAR AND WELL BELOW IT WITH RESPECT TO HORIZON. OTHER PHENOMENON A FEW DEGREES TO RIGHT OF POLAR STAR AND HIGHER ABOVE HORIZON THAN FIRST PHENOMENON. SECOND APPEARED AS STAR WHICH THEN EXPLODED FORMING GREENISH CIRCULAR CLOUD WHICH QUICKLY BECAME MISSHAPEN AND SPREAD OUT ACCROSS SKY TO DISAPPEAR. SOME REPORTED NOTHING IN CLOUDE PHASE OF SECOND PHENOMENA SHORT-LIVED BRIGHT LIGHTS.

2. WOULD APPRECIATE KNOWING EXPLANATION WHICH SMITHSONIAN MAY HAVE FOUND FOR THESE PHENOMENA.

[illegible] 3191

[redacted]

NOTE: PASSED ABOVE ADDRESSEE

BY OC/T.

UNCLASSIFIED

N4

INR-5

DEPARTMENT OF STATE

AIRGRAM

SP 16 ANT

FOR RM USE ONLY

A-208 CONFIDENTIAL

14,358-86

TO : Department of State

A000208 65

AMEMBASSY BUENOS AIRES 4 SEP 65

FROM : AmEmbassy BUENOS AIRES DATE: September 4, 1965

SUBJECT: Celestial Body Observed over Antarctica
CELESTIAL BODY OBSERVED OVER ANTARCTICA

REF : CA-1287, August 16, 1965

Attention is invited to the Department of Defense Intelligence Information Report No. 5 804 0048 65, July 16, 1965, from Buenos Aires which provided a translation of four messages received by the Argentine Navy Hydrographic Service concerning unidentified flying objects reportedly sighted in Argentina recently.

The Office of the Naval Attaché has requested information from the Argentine naval authorities related to the specific inquiry of CA-1287. It seems unlikely, however that additional information will be available until there is the summer exchange of personnel at the Argentine Antarctic stations. At the time of the personnel exchange records and other detailed data will be transferred to Buenos Aires headquarters. The Embassy will make a follow-up inquiry at that time.

As an additional item of interest, the La Plata Observatory issued a statement to the press following the public interest aroused by the Reuters dispatch of July 7 that it had no theory on the unidentified flying objects sighted in Antarctica although it would point out that the sightings occurred at the time that Echo II was orbiting over the Antarctic zone.

For the Ambassador:

Albert Mayio

Albert Mayio
Counselor of Embassy
for Commercial and Economic Affairs

CONFIDENTIAL

1965 SEP 7 PM 1 41

9-1-65

FA:AC Jr.

DCM: LJSacksio

0208

NUMBER

ARA-8

DEPARTMENT OF STATE

AIRGRAM

POL 2 MEX

FOR RM USE ONLY

A-11 UNCLASSIFIED

TO : Department of State 1965 OCT 4 AM 11 43

INFO : DEPT/ARA:MEX; MEXICO/CCA; MEXICO/POL; TIJUANA

456.80011/65 ANALYSIS & DISTRIBUTION BRANCH

FROM : American Consulate, MEXICALI DATE: September 14, 1965

SUBJECT : Political Report for August 1965

REF :

SUMMARY

1. After effects of the elections
2. San Luis Mayor has press conflict
3. French diplomats visit San Luis
4. Flying Saucers in Mexicali
5. Mexicali Mayor loses battle to move Municipal Market
6. Cotton Harvest Starts
7. All Airports to be run by one Agency
8. Restructuring of PRI in Mexicali
9. Mexicali Police Scandal
10. Press Round up

UNCLASSIFIED

FORM DS-323

Drafted by: AWFeldman/glr 9-14-65 Contents and Classification Approved by: Arthur W. Feldman

Clearances:

N6

DEPARTMENT OF STATE

AIRGRAM

FSV 7 R.S.

FOR RM USE ONLY

A-1007 UNCLASSIFIED

TO : DEPARTMENT OF STATE 1 50 5 7-29

4004/1007 66

FROM : Amembassy, OTTAWA DATE: 13 April 1966

SUBJECT: Unidentified Flying Objects

REF :

Frequent recent reportings of unidentified flying objects at various points in Canada have led to considerable interest in the subject in the House of Commons. The Library of Parliament has asked the Embassy whether any recent published reports on unidentified flying objects in the United States are available.

The Embassy has no publications which meet the needs of the Library of Parliament. It would be appreciated if the Embassy could be informed whether there are any such publications in the past two or three years. If there are any such publications available for distribution without charge, the Embassy would appreciate receiving two or three copies of each such publication. If publications are not available without charge, information concerning the source from which such publications can be purchased and the purchase price would be appreciated.

For the Charge d'Affaires a.i.:

H. Alberta Colclaser

H. Alberta Colclaser
First Secretary of Embassy

UNCLASSIFIED

FORM DS-323

ECON: HAColclaser/pb 4/13/66

ECON: CGWootton

1007 NUMBER

N7

DEPARTMENT OF STATE

23 MAY 66

AIRGRAM

SP 19

FOR RM USE ONLY

A-1354

LIMITED OFFICIAL USE

TO : Department of State
For SCI

15189-16

INFO: American Consulate, Sapporo

4 017135 4 66

FROM : American Embassy, Tokyo

DATE: May 18, 1966

SUBJECT : Sighting of object possibly originating from space vehicle.

REF : A. Dept's. CW-9986, June 14, 1962
B. Embtel 3905

The pr s item referred to in the referenced telegram is enclosed. This report irst appeared in the Japanese language Yomiuri in the evening edition of May 10, and the enclosed translation into English was published in the morning edition of May 11. As far as we are aware, no other papers have discussed the Nemuro observations.

Since the Department's instruction requesting information of this sort is quite old, we would appreciate being informed whether telegraphic notification is still desired in cases such as this where the validity of the observations seem dubious.

R. J. Wilber
REISCHAUER

Enclosure:

Press item from
The Yomiuri.

LIMITED OFFICIAL USE

FOR DEPT. USE ONLY

DS-323

Drafted by: SciAtt:RTWalker:ics 5/23/66

Approved by: SciAtt:RTW

1966 MAY 29 AM 10 14
COPY FLO-PBR

1354 NUMBER

N8

8 DEC 66

DEPARTMENT OF STATE

AIRGRAM

SP 16

FOR RM USE ONLY

A-293 — LIMITED OFFICIAL USE

15842-22

RECEIVED DEPARTMENT OF STATE

DEC 27 8 34 AM 1966

RS/AN ANALYSIS BRANCH

TO : Amembassy BUENOS AIRES, for Scientific Attache

INFO : Department of State

4 0.2 20-22.3 66

FROM : Amembassy MONTEVIDEO

DATE: December 19, 1966

SUBJECT : Unidentified Flying Objects

REF :

2 pgs

1. The Embassy has been visited by a local amateur astronomer who is rather well known in Montevideo for his long-range weather forecasts which are frequently referred to in the local press. The astronomer, Ismael REYES Febles, is director of a private observatory which he calls Antares Observatory. Reyes showed an Embassy Officer twenty color slides of the November 12 eclipse of the sun in each of which a saucerlike object was visible in the foreground in various positions throughout the cycle of the slides taken during the two-hour duration of the eclipse. Reyes said he has not shown the slides to anyone else yet and wishes to offer them to a suitable scientific agency of the United States Government. Although he mentions no price, his manner indicates that he expects remuneration other than scientific fame. The Embassy Officer who spoke to Reyes told him that there was no one in the Country Team qualified to make even a preliminary assessment of the slides but that he would bring the matter to the attention of the Scientific Attache in Buenos Aires. Reyes accepted this suggestion and said he would keep his slides secret for a reasonable period of time until he hears from the Embassy.

2. The Embassy has checked with the scientific community in Uruguay in an attempt to establish Reyes' reputation with his colleagues. The Chairman of the Astronomy Department at the University of the Republic and other reputable local scientists have reported that Reyes has no university training and is completely without standing in the scientific community. Local s[illegible]ists are generally sceptical of his work and his motives.

3. Despite the unfavorable report the Embassy has received on Reyes, the Embassy Officer who viewed the slides believes that [illegible] Reyes has indeed photographed an interesting aerial phenomenon wh[illegible] cannot be

FORM DS-323

LIMI[illegible] OFFICIAL USE

FOR DEPT USE ONLY

POL:JDS/glan:vb

Charge: J[illegible]pping

N9

DEPARTMENT OF STATE

AIRGRAM

SP IC ANG

FOR RM USE ONLY

A-53 UNCLASSIFIED

HANDLING INDICATOR

TO : Department of State (SCI)

RECEIVED

15816-144

DEPT PASS: OCSAF

DEC 23 4 05 PM 1966

INFO : Amembassy LISBON

ANALYSIS BRANCH

4307005366

FROM : Amconsul LUANDA DATE: December 23, 1966

SUBJECT: Unidentified Flying Objects Reported Over Angola

REF :

Over the weekend of December 10, the Luanda press reported the possible sighting of two "flying saucers" by the passengers and crew of a Transportes Aereos Portuguese (TAP) flight from Beira, Mozambique to Luanda on December 7. The press accounts related that a passenger, noticing what seemed to be objects "escorting" the plane when it was about 20 minutes from Luanda, informed the plane's captain, Commandant MAIA DE LOUREIRO, who looked for himself, and then reported the sighting to the Luanda Control Tower. The tower, according to Maia's statement to newsmen upon landing, informed him that there was no traffic in the area at the time. After the initial sighting, the plane's occupants reportedly continued to watch the objects until the aircraft was about one minute from touchdown in Luanda; at that point they (the objects) vanished.

The description of the objects as reported in the local press was admittedly vague, and several witnesses including Commandant Maia explained this by the high speed, maneuverability, and distance from the plane, of what they saw. Maia's only other reported comment was that he had indeed seen "two lights at the side of my aircraft." Other unattributed observations were that the objects radiated an orange glow, and that they continually manuvered in swift dives and climbs while accompanying the plane.

As a sequel to this story, the local press reported on December 13, that TAP Headquarters in Lisbon had issued a statement on the incident. Citing the company's "preliminary investigation based on the captain's report," the statement concluded that "there is no reason to consider that the phenomena observed constitute demonstrable proof of flying objects."

SUMM

UNCLASSIFIED

FOR DEPT. USE ONLY

FORM DS-323

Drafted by: :jrm

Contents and Classification Approved by: GHSumm

Clearances: Lungers

N10

DEPARTMENT OF STATE

AIRGRAM

SP 16

FOR RM USE ONLY

A-41 UNCLASSIFIED

TO : Department of State

INFO: Mexico City

FROM : Amconsul MONTERREY DATE: February 17, 1967

SUBJECT : Report of unidentified object which fell near Municipio of General Teran, N.L.

REF : Deptel 138873 and Monterrey 156

RECEIVED

FEB 23 2 02 PM 1967

27 FEB 1967

There is attached for forwarding to NASA a translation of a letter, and its enclosure, from the mayor of General Teran to the Private Secretary of the Governor of the State of Nuevo Leon regarding an unidentified object which fell near the town of General Teran on February 7, 1967.

JAMISON

Enclosures:
1. Translation of letter of Feb. 16, 1967 from the Mayor of General Teran, N.L. to the Private Secretary of the Governor of N.L.
2. Translation of the memorandum attached to above letter.

1967 FEB 23 PM 4 22

CCPYFLO-PER

UNCLASSIFIED

DS-323

TCAdams

American Embassy Tunis Report

Subject:
Unidentified Flying Objects

This page intentionally left blank.

3

NII

PAGE 01 TUNIS 01442 5042

ACTION

INFO T-01 ISO-0 PM003 CIAE-00 ODDE-0 INR-07 NSAE-00

01 USIA-06 PRS-01 SP-02 OES-03 ASA-01 /034

123403

111450 MAR 75

FM AMEMBASSY TUNIS

TO AMEMBASSY ALGIERS

INFO SECSTATE WASHDC 944

AMEMBASSY RABAT

USCINCEUR

CINCUSNAVEUR

USSIXTHFLT

R23 SPECIAL

LIMITED OFFICIAL USE TUNIS 1442

E.O. 11652: N/A

TAGS: MILI, TSPA, TS

SUBJECT: UNIDENTIFIED FLYING OBJECTS

REF: ALGIERS 638 (B) RABAT 1225

1. THERE HAVE BEEN NO REPEAT NO RECENT REPORTS IN TUNISIA OF UFO SIGHTINGS AND EMBASSY CAN SHED NO LIGHT ON PHENOMENA DESCRIBED REFTEL.

SEELYE

3

This page intentionally left blank.

Report from
Air Force Special Security Office

Subject:
(Unclassified) UFOB NEAC Area

This page intentionally left blank.

A 15

AIR FORCE SPECIAL SECURITY OFFICE
Headquarters, Northeast Air Command
APO 862 New York, N.Y.

SSO 18 July 1955

SUBJECT: (UNCLASSIFIED) UFOB NEAC Area [redacted]

TO: Director
National Security Agency
Washington 25, D.C.

Attached letter and reports forwarded to your office as requested by Commander, USAF Security Service.

LADDIE MARTIN
Captain USAF
AF SSO NEAC

If enclosures No. ______ is withdrawn (or r
attached), the classification of this correspo
ence will be down graded to *Uncl*
______ in accordance with paragra
25g ______ of AFR 205-1

55-545

AIR FORCE SPECIAL SECURITY OFFICE
Headquarters, Northeast Air Command
APO 862 New York, N.Y.

SSO 18 July 1955

SUBJECT: (UNCLASSIFIED) UFOB-NEAC Area

TO: Commander
USAF Security Service
San Antonio, Texas

1. Reference to SSO NEAC message, Cite SONEC-13, DTG 081200Z July and your message Cite COP-X5547, DTG 122311Z July 1955. Following is more detailed information on sightings in the NEAC area in June and July as requested in your message.

2. Inclosures 1, 2, 3, and 4 are detailed reports prepared by Watch Division, NEAC, on the 1 to 8 July sightings made by RB-47 crews. D/I NEAC has not correlated these sightings with any known activity.

3. Inclosures 5, 6, and 7 are detailed reports on the 7 July sighting near Harmon AFB, Nfld. Inclosure 6 contains the interrogation reports of the crews involved in the sighting. Inclosure 7, NEAC Weekly Intelligence NOFORN Digest contains a resume of the 7 July sighting and subsequent evaluation by D/I NEAC. Copies of all inclosures forwarded to DIRNSA.

LADDIE MARIN
Captain USAF
AF SSO NEAC

7 Incls
1. RB-47 sighting #1
2. RB-47 sighting #2
3. RB-47 sighting #3
4. RB-47 sighting #4
5. 64th Air Div Msg DTG 061240Z Jul 55
6. 64th Air Div Msg DTG 070100Z Jul 55
7. Weekly Intell NOFORN Digest

55-541

ITEM 1. At 2100Z on 1 June, an eastbound RB-47 experienced electronic contact with an unknown aircraft or object in the area of Devon Island. A bright return was received on the APG-32 set for 20 minutes and the warning light flashed intermittently. The RB-47's K-system recorded what was possibly the sweep of another radar nearby. Contact was lost at 2120Z. No visual contact established.

On the return westbound flight, the APG-32 again had a contact for 30 minutes duration at approximately 8,000 yards. This time no indication was registered by the warning light. No visual contact made. ****

Coordinates and times as follows:

First contact (estimate)	74°40'N-103°00'W	2100Z	1 June
Contact lost	76°15'N-81°00'W	2120Z	1 June
Second contact (estimate)	74°30'N-80°30'W	2400Z	1 June
Contact lost	74°10'N-81°00'W	0030Z	2 June

ITEM 2

At 0030Z on 4 June, a westbound RB-47 experienced electronic and visual contact with an unknown aircraft or object in the Melville Sound area. The RB-47 gun warning light was flashing intermittently and the 5 radar had a contact at 7,000 yards range. Visual contact was made by crew chief, who describes aircraft as glistening silver metallic. The aircraft was first seen low and to the rear of the RB-47. It maintained a position low and to the rear and slightly right of the RB-47. The configuration of the aircraft was obscured by contrails. The aircraft broke off contact to the north with an increase in speed. The RB-47 was at 32,000, indicating Mach .735. Although gun camera films are available, they are of such poor quality that no information can be obtained from them. The radar and visual contacts were maintained for a total of 9 minutes.****

Coordinates and times are as follows:

Contact established	74°10'N - 107°10'W	0030Z	4 June
Contact lost	73°45'N - 111°35'W	0049Z	4 June

Incl 2

ITEM 3

At 0050Z on 7 June, a westbound RB-47 enroute to Eilson AFB, Alaska, at 35,000 feet experienced electronic contact southeast of Banks Island. The RB-47's APG-32 detected an aircraft/object at 3,500 yards and at the same level as the aircraft. The scope return was small and rectangular. The pilot interpreted it to be a form of jamming. The target warning light went on and off 3 times in as many minutes. No visual or K-system contacts.****

Coordinates and times are as follows:

Contact established	71°02'N - 127°00'W	0050Z	7 June
Contact lost		0053Z	7 June (possibly later)

ITEM 4

At 2044Z on 8 June, an eastbound RB-47 flying at 30,000 feet experienced electronic and visual contact with an unknown aircraft south of Bathurst Island. The RB-47 was flying at 30,000 and indicating 450 knots. The warning light flashed and 2 indications were observed on APG-32 scope. The K-system scope showed indications of sweeping by another radar. Visual contact indicates aircraft was 5 to 10,000 feet above RB-47 and approximately 5-10 miles behind for approximately 20 minutes. The co-pilot of RB-47 states aircraft had fighter configuration. Contrail was observed as aircraft disappeared. Contrail also sighted by 2nd RB-47 approximately 80 miles behind 1st; therefore, about 70 miles behind the sighted aircraft. Distance suggests contrails remain visible for some time.****

Coordinates and times are as follows:

Contact established	74°26'N - 103°25'W	2044Z	8 June
Contact lost		2140Z	8 June

DE RJEXFT 61

Y 061240Z

FM COMDR 64TH AIR DIV DEF PEPPERRELL AFB NFLD

TO RJEDEN/COMDR AIR DEFENSE COMMAND ENT AFB COLO SPGS COLO
RJEDWP/COMDR AIR TECHNICAL INTELLIGENCE CENTER WRIGHT PAT AFB OHIO
RJEPHQ/CHIEF OF STAFF USAF WASH DC
RBEGC/COMDR EASTERN SEA FRONTIER NY NY
RJEXNE/COMDR NORTHEAST AIR COMMAND PEPPERRELL AFB NFLD
RFEMF/AIR OFFICER COMMANDING AIR DEFENSE COMMAND HQS ST HUBERTS
QUEBEC CANADA
RJCKC/AIR OFFICER COMMANDING MARITIME AIR COMMAND HALIFAX NOVA SCOTIA
RJFIC/COMDR 932ND ACWRON KEFLAVIK APRT ICELAND
RJCKNC/FLAG OFFICER ATLANTIC COAST HALIFAX NOVA SCOTIA
AF GRNC
BT

TOW DC97S ARCHIE 91 ARCHIE 29. OUT OF HARMON AFB ON REFUELING MISSION BACK TO HARMON. REPORTED SIGHTING OBJ AT 0705Z. ZC 97S ALT 20,000 FT. WEATHER CLEAR. NO CLOUDS ABOVE AIRCRAFT. SCATTER CLOUDS AT 1000 FT. LEVEL NO REFLECTION POSSIBLE FROM CLOUDS. PILOT OF ARCHIE 29 LT H H SPEER. AIRCRAFT MADE CONTACT WITH RADAR SITE. SITE PAINTED OBJ ON SCOPE AT 0707Z. HELD INTERMITTENT CONTACT FOR 49 MIN. ARCHIE 29 IN POSITION TO CLOSE ON OBJ. INSTRUCTED TO DO SO BY SITE. POSITION OF OBJ 290 DEGREES FROM STATION AT APPROXIMATELY 80 MILES. FIRST APPEARED TO BE HOVERING OBJ AT 10 OCLOCK POSITION RELATIVE TO AIRCRAFT ARCHIE 29 TRIED TO CLOSE. OBJ BEGAN TO MOVE SLOWLY N.E. 050 DEGREES T.H. RELEATIVE TO SPEED OF ARCHIE 29, 275K. PILOT OF ARCHIE 29 MAINTAINED VISUAL CONTACT WITH OBJ CALLING DIRECTION CHANGES OF OBJ TO SITE BY RADIO DIRECTION CHANGES CORROLATED EXACTLY WITH THOSE PAINTED ON SCOPE BY CONTROLLER. OBJ BEGAN CLIMBING. ARCHIE 29 PILOT LOST VISUAL CONTACT WITH OBJ'S EST ALTITUDE 40-50,000 FT. OBJ THEN ACCELERATED TO SPEED OF ESTIMATED 1600 K AND MOVED OFF IN N.E. DIRECTION PAINTED ON SCOPES. FIGHTER SCRAMBLED WHEN OBJ STARTED TO CLIMB, BENT WEAPON ON FIGHTER. NO CONTACT MADE. RADAR OR VISUAL.

PAGE 1 OF 2 PAGES

PART II

AT SAME TIMES RADAR SITE WAS ALSO PAINTING 5 SMALLER OBJ'S ON SCOPES 5-10,000 FT BELOW THE KC97'S ARCHIE 29 AND ARCHIE 91. AT 300 DEGREES. 60 MILES FROM STATION OBJ'S MOVING VERY VERY FAST. CHANGING DIRECTION AND ASIMUTH JUMPING ON AND OFF SC FORMING CIRCULAR PATTERN. CHANGING TO LINE ABREAST. TRAVELING 10-20 MILES THEN CHANGING DIRECTION CONTROLLER ESTIMATES SPEED IN EXCESS OF 1500 K. C119 AIRCRAFT ENROUTE FROM GOOSE PASSED WITHIN 5 MILES OF OBJ'S NOT KNOWN OF OBJ'S SEEN BY 119 CREW. INTERROGATION OF CREW'S OF ARCHIE 29 ARCHIE 91 AND C119 BEING CONDUCTED BY BASE INTELL OFFICER HARMON AFB COMPLETE REPORT TO BE SENT D/I 64TH A.D. FOLLOW UP MESSAGE WILL BE SENT ALL ADDRESSEE.

BT

06/1255Z JULY JEMFT

PAGE 2 OF 2 PAGES

VA6 CO1
OO CTY DE VA6 1C
O 071C

VA6 CO1
OO CTY DE VA6 1C
O 070100Z
FM COMDR 64TH AIR DIV DEF PEPPERRELL AFB NF
TO ZEN/COMDR ADC ENT AFB COL SPRINGS COL
ZEN/COMDR ATIC WRIGHT PATTERSON PATT AFB OHIO
ZEN/D/I HQ USAF WASH 25 DC
ZEN/COMDR CHIEF OF STAFF USAF WASH DC
ZEN/COMEASTSEAFRON NEW YORK NY
ZEN/CANAIRDEF ATTN OCC ST HUBERTS
ZEN/CANAIRFAX HALIFAX NOVA SCOTIA
ZEN/COMDR 932HS ACWRON KEFLAVIK APRT ICELAND
ZEN/CANFLAGLANT HALIFAX NOVA SCOTIA
ZEN/COMDR EASTERN AIR DEF COMMAND STEWART AFB NEWBURG NY
ZEN/COMDR CONTINENTAL AIR DEF COMMAND ENT AFB COL SPRINGS COL
RJEXFA/COMDR NEAC ATTN INTELLIGENCE
BT

////CONFIDENTIAL////ADOAO 167 REF ADOAO 164 AND ADOAO 165 THIS IS FOLLOW UP REPORT OF PERSONNEL STATEMENTS. THIS MSG IS IN 6 PARTS.

ADOP 7-3. THE FOLLOWING REPORT IS SUBMITTED IAW NEAC REG 200-2, DTD 13 DEC 54. THIS MSG IS IN 6 PARTS. IT INCLUDES 6 SIGNED STATEMENTS WHICH MAY CONSTITUTE A VITAL INTELLIGENCE SIGHTING.

PART 1. "I WAS AT THE CONTROLS AT THE TIME OF THE INITAIL SIGHTING OF THE OBJECT WHEN IT WAS CALLED TO THE ATTENTION OF THE CREW IN THE FORWARD PART OF THE ACFT THAT THIS PARTICULAR OBJECT WAS MAKING ERRATIC MOTION IN THE SKY. AT THIS TIME, UPON SIGHTING THE OBJECT MYSELF, I DID NOTICE THAT THE OBJECT WAS MAKING A VERY IRREGULAR ZIG ZAG PATTERN AT A POSITION OF APPROX 10 O'CLOCK LOW ON THE HORIZON. OUR ALTITUDE AT THAT TIME WAS 16,000 FEET, OUR HEADING 125 DEGREES. THE OBJECT APPEARED TO HAVE THE SAME ALTITUDE OF OUR ACFT. THE EASTERN SKY WAS JUST BEGINNING TO LIGHT UP, THE TIME BEING APPROX 0700Z, AND THE OBJECT WAS VISIBLE THROUGH WHAT LOOKED TO ME TO BE AN INVERSION LAYER. MY FIRST IMPRESSION WAS THAT IT WAS VERY LIKELY THE MORNING STAR OR VENUS AND THAT ITS ERRATIC MOTION MIGHT BE ATTRIBUTED TO REFRACTION AND OUR OWN RELATIVELY FORWARD

PAGE 1 OF 6 PAGES

MOTION. HOWEVER AT THE TIME THAT 'ENRAGE' WAS NOTIFIED AND CONFIRMED THAT THEY HAD PICKED UP 4 OR 5 TARGETS IN THIS SAME AREA, I BEGAN TO WONDER WHETHER IT WAS THE MORNING STAR OF SOMETHING ELSE. BETWEEN THE TIME OF THE INITIAL SIGHTING AND THE TIME THAT WE STARTED OUR LET-DOWN, WHICH WAS APPROXIMATELY ½ HOUR, IT IS MY ESTIMATE THAT THE OBJECT CHANGED ITS POSITION WITH RESPECT TO THE HORIZON BY APPROXIMATELY 15 DEGREES OF ELEVATION AND 5 DEGREES AZIMUTH. AFTER THIS BEFOREMENTIONED ANGLE OF ELEVATION INCREASED, THE OBJECT MOVED OUT OF THE SUSPECTED INVERSION LAYER AT WHICH TIME THE EXTENT OF ITS ERRATIC MOTION SEEMED TO DIMINISH AND ONCE AGAIN I THOUGHT IT WAS POSSIBLY THE MORNING STAR."

SIGNED LT. PAUL DAILY AO 1657967 CO-PILOT, KC-97 97TH ARS(TDY ERAFB)

PART 2. "AT 0705Z THIS DATE, I RECEIVED A CALL FROM 'ARCHIE 91' A KC-97 ENROUTE TO HARMON AFB, HIS POSITION AT THIS TIME WAS 280 DEGREES TRUE N, 80 MILES FROM THIS STATION. 'ARCHIE 91' REPORTED SIGHTING AN UNKNOWN OBJECT IN THE AIR TO THE LEFT OF HIS ACFT, SLIGHTLY ABOVE HIM, MOVING AT VERY HIGH SPEEDS. IMMEDIATELY THEREAFTER, 'ARCHIE 29' ANOTHER AIRCRAFT APPROXIMATELY 20 MILES TO THE [illegible] OF 'ARCHIE 91' REPORTED A SIGHTING OF THE SAME OBJECT IN THE SAME LOCATION.

THE OBJECT WAS DESCRIBED AS FLYING A VERY ERRATIC COURSE, MAKING SHARP CHANGES IN DIRECTION. BOTH AIRCRAFT REPORTED THAT THERE WERE NO CLOUDS WHATSOEVER NEAR THIS ALTITUDE THAT MIGHT CAUSE A REFLECTION.

APPROXIMATELY 0707Z I MADE RADAR CONTACT WITH A TARGET IN THE REPORTED POSITION, AND ALSO WITH SEVERAL OTHER TARGETS SLIGHTLY CLOSER TO THIS STATION. THE SIGHTED OBJECT APPEARED TO BE MOVING VERY SLOWLY AT THIS TIME AND EXCELLENT RAD CONTACT WAS MAINTAINED FOR SEVERAL MINUTES. THE OTHER AIRCRAFT REPORTED ALSO, THAT AT THIS TIME THE OBJECT SEEMED TO BE HOVERING. THE OTHER GROUP OF OBJECTS DETECTED ON RADAR WERE MOVING VERY FAST, SOMETIME IN A CLUSTER, THEN STRUNG OUT IN LINE FORMATION. DUE TO SLOW ANTENNA ROTATION, HIGH SPEED AND CHANGING COURSE OF THESE OBJECTS, SPEED CHECK WAS NOT AVAILABLE. ESTIMATE OF SPEED WOULD BE 1,500 KNOTS. THIS GROUP OF TARGETS WAS DETECTED BRIEFLY ON THE HEIGHT FINDER

PAGE 2 OF 6 PAGES

AT FROM 5,000 - 10,000 FEET.

I ASKED 'ARCHIE 29' IF HE WOULD ATTEMPT TO CLOSE ON THE SIGHTED OBJECT USING CAUTION, FOR A BETTER REPORT. THE PILOT OF 'ARCHIE 29', LT H.E. SPEER AGREED AND CHANGED COURSE TOWARD THE OBJECT. APPROXIMATE RANGE TO THE OBJECT WAS 40 MILES. THE OBJECT WAS STILL HOVERING IN APPROXIMATELY THE SAME LOCATION. AS THE RANGE DECREASED, THE OBJECT BEGAN SLOWLY MOVING NORTHWARD. AFTER TRAVELING NORTH ABOUT 10 MILES THE OBJECT MADE A TURN BACK SOUTH. 'ARCHIE 29' CLOSED TO WITHIN 18 MILES OF THE OBJECT, THEN IT BEGAN TO MOVE OFF IN A NORTHERLY DIRECTION AGAIN ABOUT 070 DEGREES TRUE. 'ARCHIE 29' REPORTED THAT THE OBJECT WAS MOVING NORTHWARD AND CLIMBING. HE SAID IT LOOKED LIKE HE WAS NOT CLOSING ON IT ANY LONGER. AT THIS TIME I REQUESTED PERMISSION FROM 'DIGNITY' (CONTROL CENTER) TO 'SCRAMBLE' A FIGHTER AFTER THE OBJECT. REQUEST WAS APPROVED AND 'SCRAMBLE' WAS COMPLETED. RANGE OF OBJECT BY THIS TIME WAS 330 DEGREES TRUE AT 80 MILES.

SHORTLY THEREAFTER, A MOMENTARY CONTACT WITH THE OBJECT WAS MADE ON OUR HEIGHT FINDER. ALTITUDE WAS 35,000 FEET. 'ARCHIE 29' REPORTED THAT OBJECT WAS GETTING MUCH HIGHER AND FADING FROM VIEW, DUE TO LESSENING OF SUN'S REFLECTION ON IT, AND ALSO SUN SHEAD RESTRICTED VISION.

WHEN FIGHTER BECAME AIRBORNE, OBJECT WAS 110 MILES NORTH OF THE STATION. FIGHTER PILOT, SHAKO 56, REPORTED HIS RADAR WAS INOPERATIVE WE REQUESTED HE CONTINUE MISSION FOR ATTEMPT AT VISUAL CONTACT. WHEN SHAKO 56 REACHED 'ARCHIE 29' POSITION, THE OBJECT HAD FADED FROM VIEW. BOTH AIRCRAFT WERE NOW 150 MILES NORTHEAST OF THIS STATION. AIRCRAFT WERE TRANSFERRED TO 'GRANDPA's' CONTACT MOMENTARILY, THEN TURNED BACK TOWARDS RANCH. BOTH LANDED WITHOUT FURTHER INCIDENT."

SIGNED 2ND LT. CHARLES H. DENNEY USAF SENIOR DIR 640TH ACW SQ.

AGE 3 OF 6 PAGES

PART 3. FROM THE PERIOD OF 1700 5 JULY TO 0300 6 JULY 1955, I ASSUMED THE DUTIES OF OD AT THE 640TH ACW SITE. AT APPROXIMATELY 0700Z, I WAS CALLED INTO OPERATIONS BY THE SENIOR CONTROLLER, WHO IN TURN ADVISED ME OF UNKNOWN OBJECTS BEING VIEWED ON THE RADAR. IN CHECKING ALL RADAR IN OPERATION, IT WAS FOUND THEY WERE FUNCTIONING PROPERLY.

APPROXIMATELY FROM 1 JULY 1955 TO PRESENT, PERIODICALLY THROUGH THE DAY WE HAVE HAD WEATHER INVERSIONS CAUSING OUR RADAR BEAMS TO BEND TOWARD THE GROUND. WE HAVE PERIODICALLY PICKED-UP THE SOUTH SHORE OF LABRADOR AND ANTICOSTI ISLAND DUE TO THE WEATHER INVERSION. AT THE TIME OF THE SIGHTINGS, NO SUCH RADAR PHENOMENON WAS NOTICED. ALL UNKNOWN OBJECTS SIGHTED ON THE RADAR WERE WITHIN THE CAPABILITIES OF OUR RADAR SET."

SIGNED 1ST LT. ANTHONY G. SCARPACE AO-941448 GND ELECT OFFICER, 640TH ACW SQ.

PART 4. "AT 0700Z, THE NAVIGATOR GAVE US A HEADING OF 132 DEGREES TO HARMON. JUST AFTER WE TURNED TO THAT HEADING LT. SCHRECK MADE HIS REPORT OF THE OBJECT. I LOOKED OVER WHEN HE CALLED IT IN AND THOUGHT IT WOULD HELP IF I VERIFIED WHAT HE SAW BECAUSE I SAW THE SAME THING AT APPROXIMATELY THE SAME POSITION. 'EMBRACE' IDENTIFIED BOTH OF US BY OUR PARROTT. 'EMBRACE' DETERMINED THAT I WAS APPROXIMATELY 7 MILES CLOSER TO THE OBJECT SO HE HAD ME GIVE CHASE. THE OBJECT AT THIS TIME WAS APPROXIMATELY 25,000 FEET. MY ALTITUDE WAS 18,000 FEET. THE DISTANCE WAS APPROXIMATELY 35 MILES. WHEN I FIRST SIGHTED THE OBJECT, IT MOVED TO THE LEFT IN A VERY DEFINITE MOVEMENT AND A VERY FAST MOVEMENT. THERE WERE NO SLOW INCREASE OR DECREASE IN SPEED. IT WAS A VERY DEFINITE FAST MOVEMENT. IT DID THIS TWICE. WE TURNED TO A HEADING OF 080 DEGREES AND THAT PUT THE OBJECT AT 12 O'CLOCK OUR POSITION. WHILE TRACKING THE OBJECT, IT MADE SEVERAL TURNS TO THE NORTH AND SEVERAL TURNS TO THE SOUTH WHICH WERE VERIFIED BY 'EMBRACE'. WE CLOSED OUR DISTANCE TO THE OBJECT TO APPROXIMATELY 18 MILES AND AT THAT TIME I INCREASED

PAGE 4 OF 6 PAGES

MY POWER AND TRIED TO INCREASE MY SPEED BUT DUE TO RUNNING LOW ON OIL, I HAD TO DECREASE MY POWER ONCE MORE. OUR INDICATED AIR SPEED WAS APPROXIMATELY 210 MILES PER HOUR. WHEN WE GOT WITHIN 12 MILES OF THE OBJECT, IT STARTED A DEFINITE CLIMB TO AN ESTIMATED ALTITUDE OF BETWEEN 35,000 AND 40,000 FEET. AT 0800Z, WE LOST CONTACT WITH THE OBJECT BECAUSE THE SUN WAS DIRECTLY IN FRONT OF US. AS THE SUN ROSE HIGHER, THE LIGHT BEGAN TO GET DIMMER AND DIMMER UNTIL WE COMPLETELY LOST SIGHT. IF WE HAD NOT SEEN THE MOVEMENT OF THE OBJECT, WE WOULD HAVE THOUGHT IT WAS ONLY A BRIGHT STAR, BUT THE DEFINITE HORIZONTAL AND VERTICAL MOVEMENTS OF THE OBJECT KILLED THAT BELIEF. THE OBJECT WAS SIGHTED BY AT LEAST 5 OF MY CREW MEMBERS.

AT 0625Z AS WE WERE MAKING CONTACT WITH OUR RECEIVERS, WE SIGHTED AN OBJECT WITH A BRIGHT LIGHT AT APPROXIMATELY 10,000 FEET. IT WAS BELOW US AND TO OUR LEFT HEADING IN THE SAME DIRECTION. IT WENT FORWARD TO OUR LEADER'S POSITION AND TURNED AROUND AND CAME DIRECTLY BACK FROM THE HEADING OF 090 DEGREES. MY 2ND NAVIGATOR CALLED MY ATTENTION TO THIS OBJECT, BUT I DISREGARDED IT THINKING IT WAS ANOTHER AIRCRAFT. AFTER IT PASSED OUR POSITION, WE LOST CONTACT."

SIGNED LT. HOMER E. SPEER, JR. AO-1856075 AIRCRAFT COMDR KC-97 97TH ARS (EDI ERAFB)

PART 5. "WE TURNED OFF AFTER COMPLETION OF REFUELING, THE TIME BEING APPROXIMATELY 0700Z, AND TOOK UP A HEADING FOR HARMON WHICH WAS 125 DEGREES. SHORTLY THEREAFTER 'ENRAGE' CONTROL NOTIFIED ME OF AN AIRCRAFT AT 7 MILES OFF MY LEFT WING. I VERIFIED THIS POSITION BY VISUAL CONTACT OF SAID AIRCRAFT. SHORTLY THEREAFTER, AT 10 O'CLOCK TO MY POSITION, I NOTICED A BRIGHT LIGHT. AFTER CONSULTATION WITH 3 CREW MEMBERS, WE DECIDED THE LIGHT WAS MOVING VERY RAPIDLY. I DECIDED TO CALL 'ENRAGE' CONTROL TO VERIFY THE LIGHT. I TOLD 'ENRAGE' CONTROL THAT IT WAS APPROXIMATELY 10 O'CLOCK FROM MY POSITION AT FROM 15,000 to 20,000 FEET, AN

ESTIMATED 50 MILES. I ALSO TOLD THEM IT WAS MOVING VERY RAPIDLY SOMETHING LIKE THOSE FLYING SAUCERS. SHORTLY THEREAFTER, 'ARCHIE 29' CALLED 'ENRAGE' CONTROL AND VERIFIED MY ACCOUNT OF THE OBJECT. AFTER POSITIVE IDENTIFICATION OF 'ARCHIE 29', THE DISCUSSION OF THE OBJECT WAS REFERRED TO AS TO ITS POSITION FROM 'ARCHIE 29'. I VERIFIED ALL OF 'ARCHIE 29's' INFORMATION AND CONTINUED WATCHING THE OBJECT UNTIL APPROXIMATELY 20 MILES FROM HARMON FIELD. FROM THE INFORMATION 'ENRAGE' WAS GIVING 'ARCHIE 29', IT WAS APPARENT THAT 'ARCHIE 29' WOULD MAKE ANY FURTHER REPORT ON THE OBJECT, SO I DECIDED TO LAND AT HARMON AND CLEARED FROM THE 'ENRAGE' FREQU[illegible]

SIGNED LT. ROBERT W. SCHRECK AO-775721 AIRCRAFT CMDR, KC-97 97TH ARS (TDY ERAF[illegible]

PART 6. "AFTER WE FINISHED OUR REFUELING RUN AT APPROXIMATELY 0700Z, WE RETURNED ON OUR HOMEWARD COURSE AND WE HEARD LT SCHRECK REPORT THE BRIGHT LIGHT THAT WAS MOVING VERY FAST AND ERRATICALLY. RIGHT AFTER THAT THE AIRCRAFT COMMANDER SAW THE OBJECT AND POINTED IT OUT TO ME. I COULD SEE IT. IT LOOKED LIKE A VERY BRIGHT LIGHT, LIKE A STAR. WE TURNED TOWARDS THE LIGHT AND THAT WAS THE FIRST TIME I NOTICED MOVEMENT. IT APPEARED STATIONARY FOR QUITE A WHILE AND THEN IT MOVED IN EITHER NORTHERLY OR SOUTHERLY DIRECTIONS AT DIFFERENT TIMES. EACH TIME IT MOVED, WE TURNED THE AIRCRAFT TO KEEP IT AT 12 O'CLOCK POSITION. 'ENRAGE' FIRST REPORTED WE WERE 45 MILES FROM THE OBJECT. THE SKY WAS FAIRLY LIGHT ON THE HORIZON AND BY CONTRAST THE OBJECT APPEARED VERY BRIGHT. IT WAS AT ITS BRIGHTEST WHEN 'ENRAGE' INFORMED US THAT WE WERE 18 MILES FROM IT. AFTER THE SKY WAS GETTING LIGHTER, THE OBJECT BECAME DIMMER AND MORE DIFFICULT TO SEE. ABOUT 0800Z, IT DISAPPEARED COMPLETELY."

SIGNED LT. DAVID [illegible] AO-3006255 CO-PILOT, KC-97 97TH ARS (TDY [illegible]AFB)

FOLLOW UP REPORTS WILL BE FORWARDED AS RECEIVED AT THIS SITE / / /CONFIDENTIAL/ / /

BT

07/1230Z JULY JAS

PAGE 6 OF 6 PAGES

Doc 0/2

AF FORM 112
APPROVED 1 JUNE 1948

COUNTRY	REPORT NO.	(LEAVE BLANK)
PANAMA	IR-4-58	

AIR INTELLIGENCE INFORMATION REPORT

11 70311

SUBJECT			
Unidentified Flying Object Report			
AREA REPORTED ON: PANAMA		FROM (Agency): Director of Intelligence - CAirC	
DATE OF REPORT: 18 March 1958	DATE OF INFORMATION: 9-10 March 1958		EVALUATION: B-1
PREPARED BY (Officer): Vernon D. Adams, Capt., USAF		SOURCE: Caribbean Command AOC	
REFERENCES (Control number, directive, previous report, etc., as applicable): AFR 200-2			

SUMMARY: (Enter concise summary of report. Give significance in final one-sentence paragraph. List inclosures at lower left. Begin text of report on AF Form 112—Part II.)

A number of unidentified radar tracks were observed 9-10 March 1958 by search and tracking radar located in the Canal Zone. Two tracks were investigated by aircraft with negative results.

Vernon D. Adams
VERNON D. ADAMS
Capt., USAF
Ass't. Director of Intelligence

APPROVED BY:

George Welter
GEORGE WELTER
Lt Col., USAF
Director of Intelligence

DOC-3
Doc-1 6 (2)
Doc-04

RETURN TO DOC-32

4 INCLS

1 WAC #769 (Uncl)
~~2 G-2 USARCARIB Report~~ (Conf) See pages 3-5
~~3 Log of M & I Section~~ (Uncl) see pages 5-7
4 Track Reports (Uncl)

DISTRIBUTION BY ORIGINATOR

1137

NOTE: THIS DOCUMENT CONTAINS INFORMATION AFFECTING THE NATIONAL DEFENSE OF THE UNITED STATES WITHIN THE MEANING OF THE ESPIONAGE ACT, 50 U.S.C.—31 AND 32, AS AMENDED. ITS TRANSMISSION OR THE REVELATION OF ITS CONTENTS IN ANY MANNER TO AN UNAUTHORIZED PERSON IS PROHIBITED BY LAW. IT MAY NOT BE REPRODUCED IN WHOLE OR IN PART, BY OTHER THAN UNITED STATES AIR FORCE AGENCIES, EXCEPT BY PERMISSION OF THE DIRECTOR OF INTELLIGENCE, USAF.

9 MAY 1958

(7)

(SECURITY INFORMATION when filled in)

SUPPLEMENT TO AF FORM 112

ORIGINATING AGENCY	REPORT NO.	PAGE 2 OF 7 PAGES
GAirC - Dir. of Intelligence	IR-4-58	

During the period 9 through 13 March, three unexplainable radar contacts have been made by equipment located in the Canal Zone. On two occasions, aircraft were vectored into the area by the radar sites, with negative results. Interrogation of scope operators has indicated that returns were strong and easily distinguished from cloud formations. Returns were definite when associated with clouds. Generally the tracks were triangular with speed of movement very erratic. Movement appears at times to be evasive action. The incident of 9 - 10 March was tracked by gun laying radar. During period of observation, radar maintenance personnel checked out their system thoroughly. In addition, lock was broken, however, the equipment immediately picked up target and locked on. A second tracking radar situated on Taboga Island, locked on the return. Target generally remained in same area half way between radar sites. Personnel stationed at sites reported seeing red and green lights but no noise was associated with lights Visibility was good. However, lights were visible, for only a short period. A commercial flight volunteered to investigate target. He was vectored within a hundred yards of target and reported negative sighting. Target faded out at 0208R on 10 March.

At 10:12R on 10 March, search radar reported unidentified target west of canal. A T-33 from Howard Field was sent to investigate. Negative results. Aircraft was in the immediate area of target with negative sighting. Contact with target was broken at 14:15R.

VERNON D. ADAMS
Captain, USAF
Ass't. Director of Intelligence

APPROVED BY:

George Welter
GEORGE WELTER
Lt Col., USAF
Director of Intelligence

WARNING: This document contains information affecting the national defense of the United States within the meaning of the Espionage Laws, Title 18, U. S. C., Sections 793 and 794. Its transmission or the revelation of its contents in any manner to an unauthorized person is prohibited by law. It may not be reproduced in whole or in part, by other than United States Air Force agencies, except by permission of the Director of Intelligence, USAF.

AF FORM 1 OCT 52 112a REPLACES AF FORM 112-PART II, 1 JUN 48, WHICH MAY BE USED. CLASSIFICATION (SECURITY INFORMATION when filled in)

CLASSIFICATION (SECURITY INFORMATION when filled in)

SUPPLEMENT TO AF FORM 112

ORIGINATING AGENCY	REPORT NO.	PAGE 3 OF 7 PAGES
AC OF S, G-2 USARCARIB	IR-4-58	

In accordance with Department of the Army Intelligence Collection Memorandum #200-72B-1, dated 6 August 1957, subject: "Unconventional Aircraft", the following information is submitted:

1. On 10 March 1958, Capt. Harold E. Stahlman, Operations Officer, 764th Anti-Aircraft Operations Center (AAOC), Fort Clayton Canal Zone, reported information concerning the sighting of an unidentified flying object. At 2003R, 9 March 1958, Stahlman, as Deputy Defense Commander for Anti-Aircraft Defenses, was notified at his home by the Operations Duty Officer, AAOC, that the AAOC had received a radar report of an unidentified aircraft approaching the Pacific side of the Isthmus of Panama. Stahlman arrived at the AAOC at approximately 2008R.

During the radar tracking of the first blip which appeared on the radar screen, two additional blips were observed at 2045R. The first echo was identified as a Chilean Airlines aircraft which landed at Tocumen Airport, Tocumen, Republic of Panama. The two other blips, which were not identified, indicated that the two objects were in the vicinity of Fort Kobbe, Canal Zone. A civilian aircraft in the general vicinity of the objects made a visual search of the area with negative results. The original blips were picked up by Search Radar and then transferred to the Track Radar Unit located at Flamenco Island, Fort Amador, Canal Zone. This unit was able to lock on the unidentified objects and the following information was obtained:

Number of Objects:	Two, approximately one hundred yards apart.
Duration of Radar Observation:	2003R, 9 March 1958, to 0208R, 10 March 1958.
Location of Radar:	Battery D, 764th AAA Bn, Flamenco Island
Location of Object:	LJ 2853. (Geo-Ref, Military Grid Reference System)
Prevailing Weather:	Clear, visibility unlimited; no wind reported.
Direction of Flight:	Average angle of elevation, 365°, Azimuth, 330 mils.
Manner of Flight:	Steady, slight circular path over the vicinity of Fort Kobbe, Canal Zone.
Altitude:	Varied from two to ten thousand feet. Average of seven thousand feet.

An attempt was made by members of the Radar Site, Flamenco Island, to observe the objects by searchlights. When the light touched the objects, they traveled from an altitude of two thousand feet to ten thousand feet in five to ten seconds.

AF FORM 1 OCT 52 112a REPLACES AF FORM 112-PART II, 1 JUN 48, WHICH MAY BE USED. CLASSIFICATION (SECURITY INFORMATION when filled in)

This was such a rapid movement, that the Track Radar, which was locked on target, broke the Track Lock and was unable to keep up with the ascent of the objects. As Track Radar can only be locked on a solid object, which was done in the case of the two unidentified flying objects, it was assumed that the objects were solid. The possibility that the sightings might have been weather balloons was discarded when the Air Force was contacted and stated that no balloons were in the air at that time. (F-3)

2. On 10 March 1958, Capt. Stahlman made another report concerning the sighting of an unidentified flying object by Search Radar located on Taboga Island, Republic of Panama. The following information was obtained in regard to the sightings:

Number of Objects:	One.
Duration of Radar Observation:	1012R to 1412R, 10 March 1958.
Location of Radar:	Taboga Island Radar Site.
Location of Object:	KL 1646. (Geo-Ref, Military Grid Reference System)
Prevailing Weather:	Partly cloudy.
Manner of Flight:	From an erratic to a triangular shaped flight pattern.
Altitude:	Undeterminable due to radar system used.
Speed:	Variable, from hovering to approximately one thousand miles per hour

Track Radar indicated that the object moved away from two United States Air Force jet aircraft that were approaching. At that time the speed of the object was calculated at approximately one thousand miles per hour. The use of Track Radar was terminated at 1412R.

3. On 11 March 1958, Lt. Roy M. Strom, Operations Officer, 764th AAA Bn, Fort Clayton, Canal Zone, reported information received from a Pan-American Airlines Pilot concerning an unidentified flying object. At approximately 0400R, 11 March 1958, the pilot of incoming aircraft C-509, a Pan American Airlines DC-6, observed an unidentified flying object 12 degrees North on Fox Trot route. The object appeared larger than the aircraft and was traveling in a Southeasterly direction.

CLASSIFICATION (SECURITY INFORMATION when filled in)

SUPPLEMENT TO AF FORM 112

ORIGINATING AGENCY	REPORT NO.	PAGE 5 OF 7 PAGES
AC OF S, G-2 USARCARIB	IR-4-58	

At the same time Lt. Strom reported that an unidentified flying object was picked up by Hawk Radar. The object was plotted twice at approximately 0508R heading Northwest at LK 3858. On the third plot, at 0517R, the object had moved to LK 5434 in a Southwesterly direction. Eleven minutes elapsed during the confirmation of the three plots. At 0528R, the object was sighted at LK 4303. Incoming aircraft C-509 was in the same area and Hawk Radar was asked if it was the same track that was picked up previously. The answer was negative. The object was last plotted at LJ 3254 at 0536R, still traveling in a Southwesterly direction. Radar contact was lost at that time. The size, shape, or altitude of the object could not be determined by radar. (F-6)

G2 USARCARIB COMMENT: DAICM #200-72B-1 mentioned above requires that:

"The Headquarters of the nearest Major Air Command should be notified of sightings which come to the attention of Army personnel", referenced DAICM continues, "Air Force Commanders have instructions from the Department of the Air Force which cover reporting on subject (AFR-200-2: "Unidentified Flying Objects Reporting, Short Title: UFOB)' (U)". This office continues to report information as developed.

UNCLASSIFIED (SECURITY INFORMATION when filled in) (11)

SUPPLEMENT TO AF FORM 112

ORIGINATING AGENCY	REPORT NO.	PAGE 26 OF 87 PAGES
CAirC, Director of Intel.	IR-4-58	

EXTRACTS FROM THE LOG AT MOVEMENT & IDENTIFICATION SECTION ADCC

09 March

19:59 Unknown aircraft flying Tango Route inbound. No known aircraft in area but one advised by Tocumen, WBZ BLB ATC.

20:45 Unidentified blip believed to be weather balloon picked up between Albrook and Taboga. Appears to be orbiting. No air traffic in area at all. Advised ATC of possibility of object interfering with air traffic.

20:45 Advised that a balloon had been released earlier in the evening at approximately 1830R but should be down southeast of Albrook at present time.

21:40 Tower advised P-501 (Pan American Flight) is cleared by Albrook ATC for DF instructions in order to avoid object. Flight P501 will cross the canal over Albrook.

23:45 Distance of object from D Battery (Flamingo) is 4870 yards, height 3.5 thousand feet. At the present time, a searchlight from harbor entrance control point is being used to aid in identification purposes, to be executed by one AF-Naval crash boat.

23:55 Object now at 6.0 feet moving away very rapidly to the southwest.

24:00 Radar advises that as soon as searchlight was employed, the object became evasive. Object now at 10.0 feet, 7800 yards from site. Two returns, one at 10.0 feet, other at 08.

10 March

00:44 Braniff Flight 400 reports negative sighting of object during brief investigation. Radar reported aircraft was approximately 100 yards from object.

00:55 Radar reports two targets now approximately 100 yards apart. Braniff Flight 400 landed at 0047.

02:10 Radar contact lost.

10:12 Unknown aircraft at KJ1646, speed 290K. No known aircraft in area. Check with Tocumen, Albrook, Howard, ATC & CAA. Blip very practical, has reached speed of 900K then slows to a complete stop for several minutes before moving again.

10:30 Major Davis at Howard Operations advised of UFO. He will go up and take a look.

11:20 AF 5289 (T-33) airborne to check UFO. UFO was observed

WARNING: This document contains information affecting the national defense of the United States within the meaning of the Espionage Laws, Title 18, U. S. C., Sections 793 and 794. Its transmission or the revelation of its contents in any manner to an unauthorized person is prohibited by law. It may not be reproduced in whole or in part, by other than United States Air Force Agencies, except by permission of the Director of Intelligence, USAF

AF FORM OCT 52 112a REPLACES AF FORM 112-PART II, 1 JUN 48, WHICH MAY BE USED. (SECURITY INFORMATION when filled in) 6

UNCLASSIFIED

(12)

UNCLASSIFIED
CLASSIFICATION (SECURITY INFORMATION when filled in)

SUPPLEMENT TO AF FORM 112

ORIGINATING AGENCY	REPORT NO.	PAGE	OF	PAGES
CAirC, Dir. of Intelligence	IR-4-58	7	7	

to avoid jet. As soon as jet got close, UFO appeared to move away for several miles, then stop.

12:00 Jet returned to Howard. Pilot called and advised nothing was seen. Also advised Weather Build-up. Radar advised they could distinguish UFO from clouds and build-ups.

AF FORM OCT 52 112a REPLACES AF FORM 112-PART II, 1 JUN 48, WHICH MAY BE USED. CLASSIFICATION UNCLASSIFIED (SECURITY INFORMATION when filled in) 7

TCICE952ZKZK OO ZNO ADD ZRN ZDZ ZHO DE

CRF321

OO RUEADWS

DE RUWTPFA #8016 2343929

ZNY

O 222724Z AUG 69

FM SMM MARCH AFB CALIF

TO RHCOARD/ERC OFFUTT AFB NEBR

INFO RUTWEDG/JARCC KEY WEST NAS FLA

RUSHJCS/JCS J 3 JRC WASH DC

RUCFHQA/CSAF AFXOPFR WASH DC

RUCRJCS/DIA CIIC DIAAQ 1 WASH DC

RUEADWS/DIRNSA P04 K41 CST FT MEADE MD

RUEBNXA/CINCLANT JRC NORFOLK VA

RU[illegible] CINCAFLANT LANGLEY AFB VA

3

[illegible]/[illegible] CDC GUNTER AFB ALA

[illegible]/[illegible] KEY WEST NAS FLA.

Encl 2

RUKLAAA/CINCLANTFLT NORFOLK VA

RULGARD/COMNAVBASE GUANTANAMO NAS CUBA

RULGAEA/CONCARIBASEAFRON SAN JUAN PR

RUCILBA/966AWWC SQ MCCOY AFB FLA

RUEDNKA/DIRNAVSECGRULANT WASH DC

RUWHSAA/RECHOPSDET USNS MULLER WASH DC

RULGARD/NAVSECGRUACT GUANTANAMO NAS CU

RULGARD/AAWC GUANTANAMO NAS CUBA

PAGE 2 RUWTPFA0016

RUCLEFA/DET 1 6947SSG HOMESTEAD AFB FLA

RUCLSRA/CO 6TH USASA FIELD STA HOMESTEAD AFB FLA

RUCJAAA/CINCSTRIKE MCADILL AFB FLA

RUCLEFA/4489SP SQ HOMESTEAD AFB FLA

RUWTEJA/6947SS NAS KEY WEST FLA

RUWTEJA/USAFSS TAD KEYYL AFB TEX

RUCLFOA/CCG AIR STATION MIAMI FLA

RUCBAAA/6949S SQ OPS OFFUTT AFB NEBR

RUWHNLD/CINCONAD ENT AFB COLO

RU[illegible]/CCG DIST 7 MIAMI FLA

3

O [illegible] 11 AUG 69

BT [illegible]

TO ZEN/15AF

ZEN/8AF

ZEN/SAC

ZEN/390SMW

ZEN/55SRW

BT

JOPREP JIFFY/OPREP-4/JJ0157/FOR 15AF(DI/DOR/DMAC)

8AF(DOC) 2AF(DOC) DM021/PASS TO 12SAD DO/DM

PAGE 3 RUWTPFA0016

A. REFERENCE -

A 1. BURNING PIPE

A2. B-474/B-971

A3. OPREP 1A JJ0129 OPREP 1B JJ0117 OPREP 2 JJ0153

B. FORCE:

B1. TYPE ACFT/TAIL NR/ATD/ARP BASE/ATA/ALT FLOWN/1ST TOT/LAST TOT/

PERCENT SUCCESS:

RC-135/63-9792 /21/1504Z/KOFF/22/0411Z/300-360/2117ABZ/220117Z/

100 PERCENT.

F. [illegible]IAL TACTICS: NONE

3

G. [illegible]:

G1. WEATHER: AS BRIEFED

G2. CONTRAIL START/STOP ALTITUDE: UNK

I. ROUTE OPTIONS/AIR ABORTS/DEVIATION:

1949Z(1916N) (74-40W) TO 1953Z(19-24N)(75-09W) TO 2001Z(1921N()76-15W)

TO 2016Z(1916N()78-00W) THIS WAS TO AVOID W1001(RESTRICTED AREA) DEVIATED

LEFT AT 2031Z(20-21N()79-37W)

TO 2041Z (20-57()80-48W) FOR WX AT 0059Z 21-15N 85-35W TO 0107Z

21-54N 86-12W TO 0117Z 2306N 86-32W FOR WX

K. LOSS/DAMAGE: NONE

PAGE 4 RUWTPFA0016

Z. REMARKS:

Z1. EQUIPMENT MALFUNCTION: STRIPPED NOISY SCAN BAND 4 LRU 452 OVER

LOADED CONTINUALLY AFTER 2 AND ON HALF HOURS OF OPERATION UNABLE

TO INTERCEPT SIGNALS AFTER THIS TIMETHE ALF-C DE ANTENNA BECOME INOP AFT

ER 9 HOURS OF OPERATION.

Z2. SIGHTINGS: 22/0033Z A BRIGHT WHITE LIGHT APPEARED IN THE NORTH

JUST ABOVE THE HORIZON. THE LIGHT WHICH APPEARED AS A GIANT PINWHEEL

OF A FIREWORKS DISPLAY ROSE IN ALTITUDE TO AN ANGLE OF APPROX. 40 DEG.

FROM THE HORRIZON GROWING BRIGHTER THEN WENT OUT.

3

Z3. FOOTAGE EXPOSED/CAMERA CONFIGURATION: NONE

Z4. TIME/STATION ACKNOWLEDGING:

OPS NORM 211742 HERSHEY

OPS NORM 211850 HERSHEY

OPS NORM 21016 #ERSHEY

OPS NORM 212125 HERSHEY

OPS NORM 212235 HERSHEY

OPS NORM 220107 HERSHEY

25. FIGHTER SUPPORT: NONE REQUIRE

26. ADDITIONAL REMARKS:

PERCENT SYSTEM EFFECTIVENESS: 77.9 PERCENT.

OP-3.

BT

#0816

NNNN

3

General Accounting Office Report to the Honorable Steven H. Schiff, House of Representatives

Government Records - Results of a Search for Records Concerning the 1947 Crash Near Roswell, New Mexico

This page intentionally left blank.

GAO

Report to the Honorable Steven H. Schiff, House of Representatives

GOVERNMENT RECORDS

Results of a Search for Records Concerning the 1947 Crash Near Roswell, New Mexico

NSIAD-95-187

GAO

United States
General Accounting Office
Washington, D.C. 20548

National Security and
International Affairs Division

B-262046

July 28, 1995

The Honorable Steven H. Schiff
House of Representatives

Dear Mr. Schiff:

On July 8, 1947, the Roswell Army Air Field (RAAF) public information office in Roswell, New Mexico, reported the crash and recovery of a "flying disc." Army Air Forces personnel from the RAAF's 509th Bomb Group were credited with the recovery. The following day, the press reported that the Commanding General of the U.S. Eighth Air Force, Fort Worth, Texas, announced that RAAF personnel had recovered a crashed radar-tracking (weather) balloon, not a "flying disc."

After nearly 50 years, speculation continues on what crashed at Roswell. Some observers believe that the object was of extraterrestrial origin. In the July 1994 Report of Air Force Research Regarding the Roswell Incident, the Air Force did not dispute that something happened near Roswell, but reported that the most likely source of the wreckage was from a balloon-launched classified government project designed to determine the state of Soviet nuclear weapons research. The debate on what crashed at Roswell continues.

Concerned that the Department of Defense (DOD) may not have provided you with all available information on the crash, you asked us to determine the requirements for reporting air accidents similar to the crash near Roswell and identify any government records concerning the Roswell crash.

We conducted an extensive search for government records related to the crash near Roswell. We examined a wide range of classified and unclassified documents dating from July 1947 through the 1950s. These records came from numerous organizations in New Mexico and elsewhere throughout DOD as well as the Federal Bureau of Investigation (FBI), the Central Intelligence Agency (CIA), and the National Security Council. The full scope and methodology of our work are detailed at the end of this report.

Results in Brief

In 1947, Army regulations required that air accident reports be maintained permanently. We identified four air accidents reported by the Army Air Forces in New Mexico during July 1947. All of the accidents involved military aircraft and occurred after July 8, 1947—the date the RAAF public information office first reported the crash and recovery of a "flying disc" near Roswell. The Navy reported no air accidents in New Mexico during July 1947. Air Force officials told us that according to record-keeping requirements in effect during July 1947, there was no requirement to prepare a report on the crash of a weather balloon.

In our search for records concerning the Roswell crash, we learned that some government records covering RAAF activities had been destroyed and others had not. For example, RAAF administrative records (from Mar. 1945 through Dec. 1949) and RAAF outgoing messages (from Oct. 1946 through Dec. 1949) were destroyed. The document disposition form does not indicate what organization or person destroyed the records and when or under what authority the records were destroyed.

Our search for government records concerning the Roswell crash yielded two records originating in 1947—a July 1947 history report by the combined 509th Bomb Group and RAAF and an FBI teletype message dated July 8, 1947. The 509th-RAAF report noted the recovery of a "flying disc" that was later determined by military officials to be a radar-tracking balloon. The FBI message stated that the military had reported that an object resembling a high-altitude weather balloon with a radar reflector had been recovered near Roswell.

The other government records we reviewed, including those previously withheld from the public because of security classification, and the Air Force's analysis of unidentified flying object[1] sightings from 1946 to 1953 (Project Blue Book Special Report No. 14), did not mention the crash or the recovery of an airborne object near Roswell in July 1947. Similarly, executive branch agencies' responses to our letters of inquiry produced no other government records on the Roswell crash.

Reporting Air Accidents

According to press accounts from July 1947, Army Air Forces personnel from RAAF were involved in the recovery of an airborne object near Roswell. Therefore, if an air accident report was prepared, it should have

[1]According to Air Force regulation, an unidentified flying object is an airborne object that by performance, aerodynamic characteristics, or unusual features, does not conform to known aircraft or missiles, or does not correspond to Air Force definitions of familiar or known objects or unidentified aircraft.

been prepared in accordance with Army regulations. According to an Army records management official, in 1947 Army regulations required that air accident reports be maintained permanently. An Air Force official said there was no similar requirement to report a weather balloon crash.

According to an Air Force official who has worked in the records management field since the mid-1940s, air accident reports prepared in July 1947 under Army regulations should have been transferred to Air Force custody in September 1947, when the Air Force was established as a separate service.

The Air Force Safety Agency is responsible for maintaining reports of air accidents. We examined its microfilm records to determine whether any air accidents had been reported in New Mexico during July 1947. We identified four air accidents during this time period.[2] All of the accidents involved military fighter or cargo aircraft and occurred after July 8, 1947—the date the RAAF public information office first reported the crash and recovery of a "flying disc" near Roswell. According to the Army Air Forces' Report of Major Accident, these four accidents occurred at or near the towns of Hobbs, Albuquerque, Carrizozo, and Alamogordo, New Mexico. Only one of the four accidents resulted in a fatality. The pilot died when the aircraft crashed during an attempted take-off.

Search for Records

In searching for government records on the Roswell crash, we were particularly interested in identifying and reviewing records of military units assigned to RAAF in 1947—to include the 509th Bomb Group, the 1st Air Transport Unit, the 427th Army Air Force Base Unit, and the 1395th Military Police Company (Aviation).

Document disposition forms obtained from the National Personnel Records Center in St. Louis, Missouri, indicate that in 1953, the Walker Air Force Base (formerly RAAF) records officer transferred to the Army's Kansas City records depository the histories of units stationed at Walker Air Force Base. These histories included the 509th Bomb Group and RAAF for February 1947 through October 1947; the 1st Air Transport Unit for July 1946 through June 1947; and the 427th Army Air Force Base Unit for January 1946 to February 1947. We could not locate any documentation indicating that records of the 1395th Military Police Company (Aviation)

[2]These records do not include information regarding mishaps of air vehicles belonging to civilian or other government agencies. These records also do not include mishaps involving unmanned air vehicles such as remotely piloted aircraft, low-speed cruise missiles, and most balloons.

were ever retired to the National Personnel Records Center or its predecessor depositories.

The July 1947 history for the 509th Bomb Group and RAAF stated that the RAAF public information office "was kept quite busy . . . answering inquiries on the 'flying disc,' which was reported to be in [the] possession of the 509th Bomb Group. The object turned out to be a radar tracking balloon." By his signature, the RAAF's commanding officer certified that the report represented a complete and accurate account of RAAF activities in July 1947. (Excerpts from the report are contained in app. I.)

In addition to unit history reports, we also searched for other government records on the Roswell crash. In this regard, the Chief Archivist for the National Personnel Records Center provided us with documentation indicating that (1) RAAF records such as finance and accounting, supplies, buildings and grounds, and other general administrative matters from March 1945 through December 1949 and (2) RAAF outgoing messages from October 1946 through December 1949 were destroyed. According to this official, the document disposition form did not properly indicate the authority under which the disposal action was taken. The Center's Chief Archivist stated that from his personal experience, many of the Air Force organizational records covering this time period were destroyed without entering a citation for the governing disposition authority. Our review of records control forms showing the destruction of other records—including outgoing RAAF messages for 1950—supports the Chief Archivist's viewpoint.

During our review of records at FBI headquarters, we found a July 8, 1947, teletype message from the FBI office in Dallas, Texas, to FBI headquarters and the FBI office in Cincinnati, Ohio. An FBI spokesperson confirmed the authenticity of the message.

According to the message, an Eighth Air Force headquarters official had telephonically informed the FBI's Dallas office of the recovery near Roswell of a hexagonal-shaped disc suspended from a large balloon by cable. The message further stated that the disc and balloon were being sent to Wright Field (now Wright-Patterson Air Force Base, Ohio) for examination. According to the Eighth Air Force official, the recovered object resembled a high-altitude weather balloon with a radar reflector. The message stated that no further investigation by the FBI was being conducted. (A copy of the teletype message appears in app. II.)

To follow up on the July 8th message, we reviewed microfilm abstracts of the FBI Dallas and Cincinnati office activities for July 1947. An abstract prepared by the FBI Dallas office on July 12, 1947, summarized the particulars of the July 8th message. There was no mention in the Cincinnati office abstracts of the crash or recovery of an airborne object near Roswell.

Because the FBI message reported that debris from the Roswell crash was being transported to Wright Field for examination, we attempted to determine whether military regulations existed for handling such debris. We were unable to locate any applicable regulation. As a final step, we reviewed Air Materiel Command (Wright Field) records from 1947 to 1950 for evidence of command personnel involvement in this matter. We found no records mentioning the Roswell crash or the examination by Air Materiel Command personnel of any debris recovered from the crash.

Queries to Federal Agencies Regarding Records on the Crash

We sent letters to several federal agencies asking for any government records they might have concerning the Roswell crash. In this regard, we contacted DOD, the National Security Council, the White House Office of Science and Technology Policy, the CIA, the FBI, and the Department of Energy.

The National Security Council, the White House Office of Science and Technology Policy, and the Department of Energy responded that they had no government records relating to the Roswell crash. (Copies of their responses appear in app. III, IV, and V.) The FBI, DOD, and the CIA provided the following information.

Federal Bureau of Investigation

The FBI informed us that all FBI data regarding the crash near Roswell had been processed under Freedom of Information Act (FOIA) requests previously received by the Bureau. We reviewed the FBI's FOIA material and identified the July 8, 1947, FBI teletype message discussing the recovery near Roswell of a high-altitude weather balloon with a radar reflector. (A copy of the FBI's response appears in app. VI.)

Department of Defense

DOD informed us that the U.S. Air Force report of July 1994, entitled Report of Air Force Research Regarding the Roswell Incident, represents the extent of DOD records or information concerning the Roswell crash. The Air Force report concluded that there was no dispute that something happened near Roswell in July 1947 and that all available official materials indicated the most likely source of the wreckage recovered was one of the project MOGUL balloon trains. At the time of the Roswell crash, project MOGUL was a highly classified U.S. effort to determine the state of Soviet nuclear weapons research using balloons that carried radar reflectors and acoustic sensors. (A copy of DOD's response appears in app. VII.)

Central Intelligence Agency

In March 1995, the CIA's Executive Director responded to our letter of inquiry by stating that earlier searches by the CIA for records on unidentified flying objects produced no information pertaining to the Roswell crash. The Executive Director added, however, that it was unclear whether the CIA had ever conducted a search for records specifically relating to Roswell. In the absence of such assurance, the Executive Director instructed CIA personnel to conduct a comprehensive records search for information relating to Roswell. On May 30, 1995, the CIA's Executive Director informed us that a search against the term "Roswell, New Mexico," in all CIA databases produced no CIA documents related to the crash. (A copy of CIA's response appears in app. VIII.)

Agency Comments

A draft of this report was provided to DOD for comment. DOD offered no comments or suggested changes to the report. The Chief Archivist, National Personnel Records Center offered several comments clarifying matters dealing with records management. These comments have been incorporated into the final report where appropriate.

The CIA, the Department of Energy, the FBI, the National Security Council, and the White House Office of Science and Technology Policy also received excerpts from the report discussing the activities of their respective agencies. They had no substantive comments and made no suggested changes to the report.

Scope and Methodology

To determine the requirements for reporting air accidents in 1947, we interviewed military service records management officials, reviewed military record-keeping regulations in effect during this time period, and examined Army Air Forces and Navy air accident reports.

We also sought to identify any government records related to the Roswell crash. In this regard, we visited and reviewed records at the locations listed in table 1.

Table 1: Locations Visited and Records Reviewed

Locations visited	Records reviewed
National Archives, Washington, D.C.	Air Force papers on unidentified flying objects Army Counterintelligence Corps historical files, 1947-49
National Archives II, College Park, Md.	Project Blue Book Special Report No. 14 National Security Council meeting minutes, 1947-48
National Archives, National Record Center, Suitland, Md.	Army Inspector General reports, 1947-58 Army staff intelligence correspondence, 1947-56 Headquarters Army Air Force message traffic, 1947-54 Army Air Force and Air Materiel Command (Wright Field) research and development files, 1947-50
National Personnel Records Center, St. Louis, Mo.	Morning reports for RAAF units, July 1947 Eighth Air Force messages, 1947-50 Eighth Air Force correspondence, 1947-51 Eighth Air Force weekly activity summaries, July 1947 Service records of key personnel assigned to RAAF, 1947 Project Sign[a] investigative reports, 1948 Army Adjutant General correspondence, 1947-49 Missile test firing reports at White Sands, N. Mex., 1947-54
Department of the Air Force, Washington, D.C.	Current and past records management regulations Report of Air Force Research Regarding the Roswell Incident, July 1994
Department of the Army, Washington, D.C.	Current and past records management regulations
Department of the Navy, Washington, D.C.	Air accident reports, July 1947
Air Force Safety Agency, Kirtland Air Force Base, N. Mex.	Air accident reports, July 1947
Air Force History Support Office, Bolling Air Force Base, Washington, D.C.	509th Bomb Group and RAAF monthly histories, July and August 1947
National Security Agency, Fort Meade, Md.	FOIA records, Citizens Against UFO Secrecy
Military History Institute, Army War College, Carlisle, Pa.	Army Counterintelligence Corps reports, 1947
Army Central Security Facility, Fort Meade, Md.	Army Counterintelligence Corps reports, 1947
Central Intelligence Agency, Langley, Va.	Scientific Advisory Panel on Unidentified Flying Objects (Robertson Panel) report FOIA records, Ground Saucer Watch, Inc.
Federal Bureau of Investigation, Washington, D.C.	FOIA records on unidentified flying objects
National Atomic Museum, Kirtland Air Force Base, N. Mex.	509th Bomb Group historical information, 1947 RAAF base newspaper Atomic Blast, July and August 1947

[a]Project Sign was the predecessor to Project Blue Book.

Our search of government records was complicated by the fact that some records we wanted to review were missing and there was not always an explanation. Further, the records management regulations for the retention and disposition of records were unclear or changing during the period we reviewed.

We also queried the National Security Council, the White House Office of Science and Technology Policy, the Department of Energy, the FBI, DOD, and the CIA to determine what government records they have on the Roswell crash. We did not independently verify the information provided to us in their written responses.

In addition to physically examining government records, we contacted the following federal activities to determine whether they had any information about the Roswell crash:

- Air Force Historical Research Agency, Maxwell Air Force Base, Alabama;
- Air Force Aeronautical Systems Center, Wright-Patterson Air Force Base, Ohio;
- Army Center for Military History, Washington, D.C.; and
- 509th Bomb Wing, Office of the Historian, Whiteman Air Force Base, Missouri.

We conducted our review from March 1994 to June 1995 in accordance with generally accepted government auditing standards.

Unless you publicly announce its contents earlier, we plan no further distribution of this report until 30 days after its issue date. At that time, we will make copies available to other interested parties upon request.

If you or your staff have any questions about this report, please call me on (202) 512-3504. A major contributor to this report is Gary K. Weeter, Assistant Director.

Sincerely yours,

Richard Davis

Richard Davis
Director, National Security
Analysis

Contents

Abbreviations

CIA	Central Intelligence Agency
DOD	Department of Defense
FBI	Federal Bureau of Investigation
FOIA	Freedom of Information Act
RAAF	Roswell Army Air Field

Combined History for July 1947

COMBINED HISTORY

509TH BOMB GROUP
AND
ROSWELL ARMY AIR FIELD

1 JULY 1947 THROUGH 31 JULY 1947

RESTRICTED

The other three briefings were those which were given to the VIP and a simulated briefing to a large group of Air Scouts representing all of the troops in New Mexico which was given on 1? July 1947.

Several small projects were completed during the month including signs on all the office doors, a building directory, and a world situation map which is maintained on a day-to-day basis.

The Historical Section of S-2 has been seriously handicapped by the removal of the regular stenographer with the reduction in force.

Due to the fact that the quality of the department reports has in general been so inadequate, lectures are being prepared to be given early in August to properly train the liaison representatives of each department.

The Office of Public Information was kept quite busy during the month answering inquiries on the "flying disc", which was reported to be in possession of the 509th Bomb Group. The object turned out to be a radar tracking balloon.

The main project of the month was making all arrangements for a successful Air Force Day. Lt. Colonel Oliver LaFarge, Air Reserve Corps, at Santa Fe, made arrangements for Colonel Blanchard to visit the Governor of New Mexico and ask him to declare Air Force Day in New Mexico on 1 August.

RESTRICTED
39

FBI Teletype Message Dated July 8, 1947

TELETYPE

FBI DALLAS 7-8-47 6-17 PM [redacted]

DIRECTOR ~~AND~~ SAC, CINCINNATI URGENT [redacted]

FLYING DISC, INFORMATION CONCERNING. [redacted] HEADQUARTERS EIGHTH AIR FORCE, TELEPHONICALLY ADVISED THIS OFFICE THAT AN OBJECT PURPORTING TO BE A FLYING DISC WAS RE COVERED NEAR ROSWELL, NEW MEXICO, THIS DATE. THE DISC IS HEXAGONAL IN SHAPE AND WAS SUSPENDED FROM A BALLON BY CABLE, WHICH BALLON WAS APPROXIMATELY TWENTY FEET IN DIAMETER. [redacted] FURTHER ADVISED THAT THE OBJECT FOUND RESEMBLES A HIGH ALTITUDE WEATHER BALLOON WITH A RADAR REFLECTOR, BUT THAT TELEPHONIC CONVERSATION BETWEEN THEIR OFFICE AND WRIGHT FIELD HAD NOT [redacted] BORNE OUT THIS BELIEF. DISC AND BALLOON BEING TRANSPORTED TO WRIGHT FIELD BY SPECIAL PLANE FOR EXAMINAT INFORMATION PROVIDED THIS OFFICE BECAUSE OF NATIONAL INTEREST IN CASE . [redacted] AND FACT THAT NATIONAL BROADCASTING COMPANY, ASSOCIATED PRESS, AND OTHERS ATTEMPTING TO BREAK STORY OF LOCATION OF DISC TODAY. [redacted] ADVISED WOULD REQUEST WRIGHT FIELD TO ADVISE CINCINNATI OFFICE RESULTS OF EXAMINATION. NO FURTHER INVESTIGATION BEING CONDUCTED.

b7C

WYLY

RECORDED 62-83894-18 F B I

END

JUL 22 1947

EX-29

CXXXX ACK IN ORDER

UA 92 FBI CI MJW

b7C

BPI HS

8-36 PM O

8-22 PM OK FBI WASH D

OK FBI CI

[redacted]

Comments From the National Security Council

9K 2948

NATIONAL SECURITY COUNCIL
WASHINGTON, D.C. 20506

April 28, 1995

MEMORANDUM FOR MR. JOSEPH E. KELLEY
Director-in-Charge, International Affairs Issues
General Accounting Office

SUBJECT: Request for NSC Records

I am responding to your April 12, 1995, request for information or NSC records related to the crash of an airborne object near Roswell, New Mexico in July 1947. The NSC has no records or information related to the incident at Roswell.

For information about any government records that may document the crash at Roswell, we suggest you contact the National Archives, Textual Reference Division, 8601 Adelphi Road, College Park, Maryland 20740. Their telephone number is (301) 713-7230.

Andrew D. Sens
Executive Secretary

Comments From the Office of Science and Technology Policy, Executive Office of the President

EXECUTIVE OFFICE OF THE PRESIDENT
OFFICE OF SCIENCE AND TECHNOLOGY POLICY
WASHINGTON, D.C. 20500

April 26, 1995

Dear Mr. Hunt:

In response to your recent inquiry of April 12, 1995. The Office of Science and Technology Policy reviewed its records regarding the Roswell Incident. OSTP has no direct knowledge of what occurred at Roswell and no records, except for the information I received from the Air Force.

I look forward to receiving the GAO report.

Sincerely,

John H. Gibbons
Director

Enclosures
As stated

Mr. William Hunt
Director, Federal Management Issues
United States Government Accounting Office
Washington, DC 20548

Comments From the Department of Energy

Department of Energy
Washington, DC 20585

June 8, 1995

Richard Davis
Director, National Security
Analysis
General Accounting Office
Washington, D.C. 20538

Dear Mr. Davis:

This is in response to your request for records related to the crash of an airborne object near Roswell, New Mexico, in July 1947. We conducted an in-depth search for documents related to the crash and have found no such documents.

If you have any questions regarding this matter, please do not hesitate to contact Barry Uhlig of my staff on 202-586-1910.

Sincerely,

Joseph F. Vivona
Chief Financial Officer

Comments From the Federal Bureau of Investigation

U.S. Department of Justice

Federal Bureau of Investigation

Washington, D. C. 20535

April 24, 1995

Mr. Richard Davis
Director
Attn: Gary K. Weeter
National Security Analysis
General Accounting Office
Washington, D. C.

Dear Mr. Davis:

This is in response to a letter dated April 7, 1995, from Norman J. Rabkin, Director, Administration of Justice Issues, General Accounting Office, to John E. Collingwood, Inspector in Charge, Office of Public and Congressional Affairs, FBI, regarding government records concerning the crash of an airborne object near Roswell, New Mexico, in July 1947 (Code 701034).

A search of FBI indices for information relating to the crash of an airborne object near Roswell, New Mexico, in 1947, determined that all FBI data concerning the incident has been processed under the provisions of the Freedom of Information Act (FOIA) and is available for review in our FOIA Reading Room. A copy of the document forwarded to me by Gary Weeter is among the documents in the Reading Room. If your staff wishes to review the material, please call Margaret Tremblay, a member of my staff, at least 48 hours in advance of the desired appointment. She can be reached on 324-5292.

Sincerely yours,

Swanson D. Carter
Supervisory Special Agent
Office of Public and
Congressional Affairs

Comments From the Department of Defense

INSPECTOR GENERAL
DEPARTMENT OF DEFENSE
400 ARMY NAVY DRIVE
ARLINGTON, VIRGINIA 22202-2884

Analysis
and Followup

MAY - 4 1995

Mr. Richard Davis
Director, National Security Analysis
U.S. General Accounting Office
441 G Street N.W.
Room 5025
Washington, D.C. 20548

Dear Mr. Davis:

The Department of the Air Force July 1994 report is the DoD response to questions posed in your April 12 letter related to GAO Code 701034.

If you have any questions, please contact my action officer, Pattie Cirino, at 703-604-9631. If she is not available, please contact Ms. Marlene L. Scales at 703-604-9626.

Kathryn M. Truex
Deputy Assistant Inspector General
for GAO Report Analysis

Enclosure

Appendix VIII

Comments From the Central Intelligence Agency

Central Intelligence Agency

Washington, D.C. 20505

30 May 1995

Mr. Richard Davis
Director, National Security Analysis
US General Accounting Office
Washington, D.C. 20548

Dear Mr. Davis,

In a letter dated 30 March 1995, this Agency advised you that it would conduct a comprehensive record search to aid in the completion of your investigation of the crash of an airborne object near Roswell, New Mexico, in July 1947. In accordance with your request, we have searched all of our data bases against the terms "Project Mogul" and "Roswell, New Mexico."

The search did not yield any documents related to either of these terms. Therefore, this Agency has no information, beyond those records already reviewed by Mr. Gary Weeter of your staff, relevant to your investigation.

Sincerely,

Nora Slatkin

Nora Slatkin
Executive Director

This page intentionally left blank.

* Memorandum and Order - Citizens Against UFO Secrecy v. National Security Agency

(Civil Action No. 80-1562)

This page intentionally left blank.

IN THE UNITED STATES DISTRICT COURT
FOR THE DISTRICT OF COLUMBIA

CITIZENS AGAINST UFO SECRECY, Plaintiff,	)	
v.	)	Civil Action No. 80-1562
NATIONAL SECURITY AGENCY, Defendant.	)	

MEMORANDUM AND ORDER

This Freedom of Information Act case is before the Court on defendant's motion for summary judgment. Plaintiff is seeking all documents in defendant's possession relating to UFOs and UFO phenomena. Defendant has provided some material but has withheld other material pursuant to various FOIA exemptions, see 5 U.S.C. § 552(b) (1976). The bulk of the material withheld consists of communications intelligence reports, which defendant asserts are protected by Exemptions 1 and 3 of the Freedom of Information Act. Four documents at issue are not communications intelligence reports; defendant has withheld one of these documents in its entirety, and portions of three others, pursuant to these exemptions, and exemptions 5 and 6.

The Court first carefully reviewed the public affidavit of National Security Agency official Eugene Yeates and then, after receiving plaintiff's opposition, examined personally a top secret affidavit from Yeates, submitted by defendant in camera. The public affidavit provides a general statement of the defendant's position that is strongly bolstered by the extremely detailed, 21-page in camera filing. On the basis of these affidavits, the Court finds that the claimed exemptions have been properly and conscientiously applied.

The communications intelligence reports clearly relate to the most sensitive activities of the defendant and thus fall squarely within the protection of Public Law 86-36, 73 Stat. 63 (1959). The in camera affidavit provides the kind of detail approved in Hayden & Fonda v. National Security Agency, 608 F.2d 1381, 1389-91 (D.C. Cir. 1979), cert. denied, 48 U.S.L.W. 3730 (U.S., May 12, 1980). The Court finds that release of this material could seriously jeopardize the work of the agency and the security of the United States. Under the standards set forth in Hayden & Fonda, the claim of Section (b)(3) exemption must be granted in its entirety. No consideration needs to be given the additional claim for protection under Exemption 1. The standards of Public Law 86-36 have been met.

The Court also finds that the affidavits support nondisclosure for the four documents that are not intelligence reports. Defendant has provided plaintiff with much of the contents of these documents and also has provided a description of both the documents and the deletions. The various claims under Exemptions 1, 3, 5 and 6 as to these documents are proper. The withheld portions either are not responsive to plaintiff's request or are properly exempted.

Throughout the Court's review of this material, the Court has been aware of the public interest in the issue of UFOs and the need to balance that interest against the agency's need for secrecy. The in camera affidavit presents factual considerations which aided the Court in determining that the public interest in disclosure is far outweighed by the sensitive nature of the materials and the obvious effect on national security their release may well entail.

The Court has been mindful of the Court of Appeals' view on in camera review of documents in FOIA cases, see Allen v. CIA, No. 80-1380 (D.C. Cir., filed Nov. 12, 1980), but the Court finds in its discretion that the public and in camera affidavits submitted here are amply sufficient to make such review of the documents unnecessary.

For the foregoing reasons, defendant's motion for summary judgment is granted. The case is dismissed.

SO ORDERED.

Gerhard A. Gesell
UNITED STATES DISTRICT JUDGE

November 18, 1980.

This page intentionally left blank.

Communication with Extraterrestrial Intelligence

by Lambros D. Callimahos

This page intentionally left blank.

Communication with Extraterrestrial Intelligence[1]

BY LAMBROS D. CALLIMAHOS

Unclassified

We are not alone in the universe. A few years ago, this notion seemed farfetched; today, the existence of extraterrestrial intelligence is taken for granted by most scientists. Sir Bernard Lovell, one of the world's leading radio astronomers, has calculated that, even allowing for a margin of error of 5000%, there must be in our own galaxy about 100 million stars which have planets of the right chemistry, dimensions, and temperature to support organic evolution. If we consider that our own galaxy, the Milky Way, is but one of at least a billion other galaxies similar to ours in the observable universe, the number of stars that could support some form of life is, to reach for a word, astronomical. As to advanced (by miserable earth standards) forms of life, Dr. Frank D. Drake of the National Radio Astronomy Observatory at Green Bank, West Virginia, has stated that, putting all our knowledge together, the number of civilizations which could have arisen by now is about one billion. The next question is, "Where *is* everybody?"

The nearest neighbor to our solar system is Alpha Centauri, only 4.3 light years away; but, according to Dr. Su-Shu Huang of the National Aeronautics and Space Administration, its planetary system is probably too young for the emergence of life. Two other heavenly friends, Epsilon Eridani and Tau Ceti, about 11 light years away, are stronger contenders for harboring life. Nevertheless, if superior civilizations are abundant, the nearest would probably be at least 100 light years away; therefore, it would take 200 years for a reply to be forthcoming, a small matter of seven generations. This should, however, make little difference to us, in view of the enormous potential gain from our contact with a superior civilization. Unless we're terribly conceited (a very unscientific demeanor), we must assume that the "others" are far more advanced than we are. Even a 50-year gap would be tremendous; a 500-year gap staggers the imagination, and as

[1] The substance of this article was presented at a panel discussion of the same title during the 1965 IEEE Conference on Military Electronics held in Washington, D. C., on 23 September 1965. Besides the author as cryptologist, the other members of the panel were Dr. Paul Garvin, linguist; Dr. John C. Lilly, delphinologist; Dr. William O. Davis, physicist; and Fr. Francis J. Heyden, S. J., astronomer. The moderator was Dr. Harold Wooster, Director of Information Services of the Air Force Office of Scientific Research.

for a 5000-year gap . . . (By the way, if they are as much as 50 years *behind* us, forget it!) It is quite possible that "others" have satellite probes in space, retransmitting to "them" anything that sounds non-random to the probe. But they have probably called us several thousand years ago, and are waiting for an answer; or worse yet, they have given up; or, more probably, they have reached such impressive technological advances that they have destroyed themselves.[2]

Epsilon Eridani and Tau Ceti were the targets on which Dr. Drake focussed his attention in the spring of 1960 in Project Ozma, an attempt to detect possible intelligent signals from outer space. The frequency selected for listening was 1420.405752 megacycles per second, or a wave length of 21 cm. This particular frequency, postulated independently by two professors on the faculty of Cornell University, Giuseppe Cocconi and Philip Morrison, happens to be the radiation frequency of atomic or free hydrogen which permeates space in great clouds; moreover, this frequency is within the range of radio frequencies able to pass through the earth's atmosphere. Presumably, the significance of this frequency would be known to other intelligent beings in the universe who understand radio theory. We're still talking about radio waves as the communication medium; other possible media might be masers, lasers, or the as-yet undiscovered and unnamed "rasers." A technology superior to ours might even have learned how to modulate a beam of neutrinos (weightless, uncharged particles that physicists on earth find it difficult even to detect); if so, "they" may have to wait a century or two before we learn how to build a neutrino receiver.

If another civilization were trying to establish communication with us, it would first embark on attention-getting signals of such a nature that we could distinguish them from random cosmic noise; once we receive a recognizable signal, we have a good chance of understanding the message. For example, they could start with trains of signals corresponding to the natural numbers 1, 2, 3, . . . , followed perhaps by prime numbers. They might continue with equal-length extended signals consisting of start and stop impulses, with occasional pulses in

---

[2] In this connection, Professor Iosif Shklovsky, Russia's greatest radio astronomer, has the following to say in the September 1965 issue of *Soviet Life:*

"Profound crises lie in wait for a developing civilization and one of them may well prove fatal. We are already familiar with several such critical [situations]:

(a) Self-destruction as a result of a thermonuclear catastrophe or some other discovery which may have unpredictable and uncontrollable consequences.
(b) Genetic danger.
(c) Overproduction of information.
(d) Restricted capacity of the individual's brain which can lead to excessive specialization, with consequent dangers of degeneration.
(e) A crisis precipitated by the creation of artificial intelligent beings."

between; when these signals are a
would show a circle, the Pythagor
design. These attention-getting s
"language lessons," interspersed wi
help bring us up to the level of our s

It may be assumed that the s
possessed by all higher forms of li
could thus be greatly simplified
representation such as that of a te
held at Green Bank in 1961 to discu
with other planets, one of the part
up a hypothetical message on th
consisting of 1271 binary digits or
1271 has but two prime factors, 3
to write out the message in raster
in 31 lines of 41 bits each; the latte
ness in the patterns disclosed, in
dimensions. In Fig. 2 is the writ
binary 1's have been replaced by a
Now for its interpretation.

There are dots at the four co
points, marking the outlines of th
representation of the sun; directl
representing 8 planets, identified
their left, preceded by a binary
legged beings illustrated are obvi
hand of the male figure points to
parently reside. At the top of th
tions of hydrogen, carbon, and
chemical structure of life on their
third planet there emerges a wavy
water; the representation of a fish
and therefore have space travel.
to a six (preceded by the usual b
there are six fingers on each har
their number system is probably
female figure may be seen a bracke
binary form (preceded by a binar
are 11 units high. A reasonable
cm., the wave length of the trans.
tall, which should be all right for a

In 1952 the British mathemat
address before the British Interpl
or First Steps in Celestial Syntax.

', if they are as much as 50 years
sible that "others" have satellite
nem" anything that sounds non-
ave probably called us several
for an answer; or worse yet, they
ey have reached such impressive
e destroyed themselves.[2]
the targets on which Dr. Drake
g of 1960 in Project Ozma, an
signals from outer space. The
is 1420.405752 megacycles per
This particular frequency, pos-
ssors on the faculty of Cornell
ilip Morrison, happens to be the
hydrogen which permeates space
quency is within the range of
h the earth's atmosphere. Pre-
uency would be known to other
understand radio theory. We're
e communication medium; other
, or the as yet undiscovered and
perior to ours might even have
neutrinos (weightless, uncharged
it difficult even to detect); if so,
two before we learn how to build

o establish communication with
-getting signals of such a nature
random cosmic noise; once we
a good chance of understanding
ild start with trains of signals
1, 2, 3, . . . , followed perhaps
inue with equal-length extended
pulses, with occasional pulses in

ky, Russia's greatest radio astronomer,
ue of *Soviet Life:*
ilization and one of them may well prove
critical [situations]:
ear catastrophe or some other discovery
rollable consequences.

n which can lead to excessive specializa-
ion.
icial intelligent beings."

between; when these signals are aligned flush over one another, they would show a circle, the Pythagorean Theorem, or similar geometric design. These attention-getting signals would be followed by early "language lessons," interspersed with items of technical information to help bring us up to the level of our superiors, "them."

It may be assumed that the sense of sight, or an equivalent, is possessed by all higher forms of life; the problems of communication could thus be greatly simplified through the medium of a "raster" representation such as that of a television screen. After a conference held at Green Bank in 1961 to discuss the possibility of communication with other planets, one of the participants, Bernard M. Oliver, made up a hypothetical message on the raster principle. The message, consisting of 1271 binary digits or "bits," is shown in Fig. 1. Since 1271 has but two prime factors, 31 and 41, we would naturally be led to write out the message in raster form, in 41 lines of 31 bits each, or in 31 lines of 41 bits each; the latter case reveals a greater nonrandomness in the patterns disclosed, indicating that these are the correct dimensions. In Fig. 2 is the write-out of the message, in which the binary 1's have been replaced by a dot and the 0's left as blank spaces. Now for its interpretation.

There are dots at the four corners of the pictogram as reference points, marking the outlines of the rectangle. At the upper left is a representation of the sun; directly underneath in a column are dots representing 8 planets, identified by the appropriate binary coding to their left, preceded by a binary point as a marker. The erect, two-legged beings illustrated are obviously bisexual and mammalian; one hand of the male figure points to the fourth planet where they apparently reside. At the top of the pictogram may be seen representations of hydrogen, carbon, and oxygen atoms, indicating that the chemical structure of life on their planet is similar to ours. From the third planet there emerges a wavy line, showing that it is covered with water; the representation of a fish shows that they must have visited us and therefore have space travel. One hand of the female figure points to a six (preceded by the usual binary point), perhaps implying that there are six fingers on each hand; we could therefore assume that their number system is probably to the base 12. At the right of the female figure may be seen a bracket, in the middle of which is eleven in binary form (preceded by a binary point): this implies that the beings are 11 units high. A reasonable interpretation is that the unit is 21 cm., the wave length of the transmission, making them about 7½ feet tall, which should be all right for average Martians.

In 1952 the British mathematician Lancelot Hogben delivered an address before the British Interplanetary Society entitled "Astraglossa, or First Steps in Celestial Syntax." Hogben pointed out that *number*

```
1000000000000000000000000000000000000000
1000011100000000001000001000000100000100
0000010001000000000000000000000000000000
0000001000100001000000100000000000010000
0000000100010000010000100000000001000100
0100000001110000000000001000000000000001
0000000000000000000000000000000000000000
0000000000000000000000001000001000000100
0001000011000100000000001100000000000000
0000000000000000000110000110000011000001
1101101101001100001100101100101100100100
1001001001010100100100001100001100001100
0011000011000000000100000000000111110100
0000000000000000000100000000000100000100
0000000001011011100100000000000001111101
0000000000000000000000000000100000000000
0000001000100111000000000000001010000000
0000001010010000110010101110010100000000
0000000101001000010000000000100100000000
0000000001001000001000000000001111100000
0000000011111000000111010100000000101010
0000000010101000000010000000000010001010
0000000010100100000000000000000001000100
1000100010011011001110110110100000100010
0010101010001000100000000000000000010001
0001001001000100010000000010000000000111
0000011111000001110000000001111101000000101
0100000101000001000100000010000000000100
0001000011100001000001000001000000000010
0000100010001000100000100000110001100001
0000010001000100010000000100000110000000001
1000001101100011011000000110011 1
```

Fig. 1.

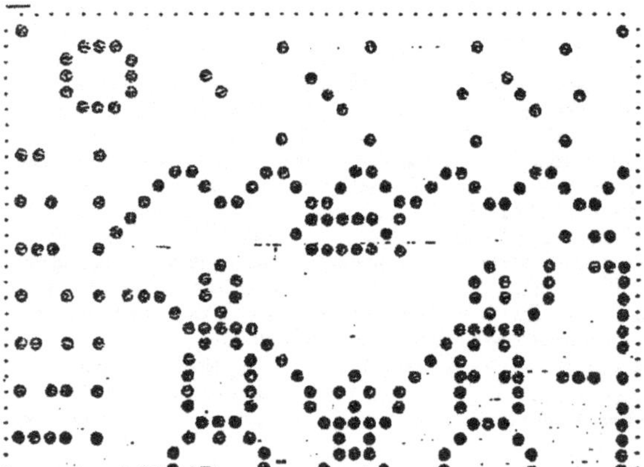

Fig. 2.

is the most universal concept for
intelligent beings; therefore, mat
steps in extraterrestrial communi
could transmit pulses representin
"radioglyphs" representing "+"
later carried out the basic idea
shapes to represent elementary
different approach was develope
Mathematics at the University
book entitled "Lincos: Design of
"Lincos," an acronym of "lingu
munication of ideas through sym
of those who have taken the trou
is too difficult. After all, the o
across to another party, whose
different from our own. In othe
an "inverse cryptography," or
designed, not to hide meaning, b
prehend. Cleverness on the par
factor, not reliance on ingenu
cryptographer—somehow, this t
make his meaning clear to the r
possess a cosmic equivalent of the

As an illustration of how much
minimum of material, and as an e
let us consider a message I have d
expect of an initial communica
shown a series of transmissions v
inhabited planet, many light yea
are representations for the 32 dif
or distinctive pulse shapes) heard
The punctuation marks are not p
different time lapses: adjacent s
unit) between them; a space be
(2 units); commas, semicolons, an
16 units, respectively. Betwee
reference purposes) there is a tim

The first transmission, (1), is
different symbols which will be u
mission (2) is the clear implicati

[3] The Rosetta Stone is a piece of blac
of the Nile, bearing a bilingual inscription
and Greek) with which Jean François Ch
Egyptian hieroglyphs.

```
0000000000000000000
0001000000010000100
0000000000000000000
0100000000000010000
0001000000010000100
0000010000000000001
0000000000000000000
0001000001000000100
0001100000000000000
10000110000110000110
0101100101100100100
0001100001100001100
0000000000111110100
0000000000100000100
0000000000001111101
0000000100000000000
0000000101000000000
0101110010100000000
0000000100100000000
0000000001111100000
1010100000010101000
0000000000010001010
0000000000001000100
0110110100000100010
0000000000000010001
0001000000000000111
0000111110100000101
0000010000000000100
1000001000000000010
0100000110001100001
0010000011000000001
0001100111
```

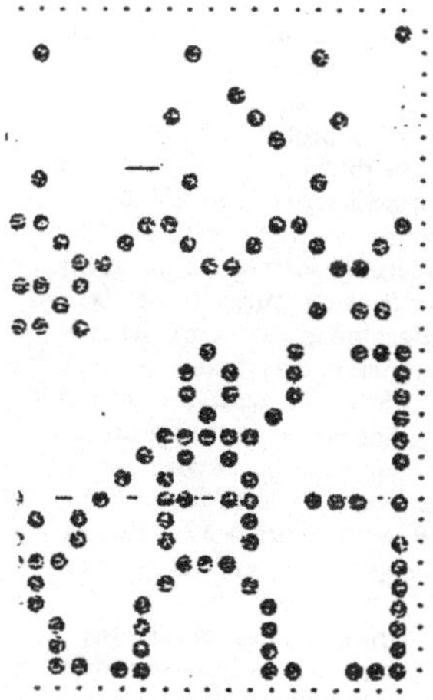

is the most universal concept for establishing communication between intelligent beings; therefore, mathematics forms the basis for the first steps in extraterrestrial communication. He then illustrated how he could transmit pulses representing integers, and distinctive signals or "radioglyphs" representing "+", "−", "=", and so on. Morrison later carried out the basic idea a little further, using different pulse shapes to represent elementary mathematical symbols. An entirely different approach was developed by Hans Freudenthal, Professor of Mathematics at the University of Utrecht, who in 1960 published a book entitled "Lincos: Design of a Language for Cosmic Intercourse." "Lincos," an acronym of "lingua cosmica," tries to establish a communication of ideas through symbolic logic, but the general consensus of those who have taken the trouble to study his book is that his plan is too difficult. After all, the object of the exercise is getting ideas across to another party, whose thinking processes may be entirely different from our own. In other words, what we need to develop is an "inverse cryptography," or communication symbolism specially designed, not to hide meaning, but to be as *easy* as possible to comprehend. Cleverness on the part of the *sender* is then the important factor, not reliance on ingenuity of the recipient. The inverse cryptographer—somehow, this term doesn't sound quite right—must make his meaning clear to the recipient, even if the latter does not possess a cosmic equivalent of the Rosetta Stone.[3]

As an illustration of how much information could be conveyed with a minimum of material, and as an example of facile inverse cryptography, let us consider a message I have devised to be typical of what we might expect of an initial communication from outer space. In Fig. 3 is shown a series of transmissions which could have come from another inhabited planet, many light years away. The 32 arbitrary symbols are representations for the 32 different signals (combinations of beeps, or distinctive pulse shapes) heard on a frequency of 1420.4 megacycles. The punctuation marks are not part of the message, but here represent different time lapses: adjacent symbols are sent with a short pause (1 unit) between them; a space between symbols means a longer pause (2 units); commas, semicolons, and periods indicate pauses of 4, 8, and 16 units, respectively. Between transmissions (numbered here for reference purposes) there is a time lapse of 32 units.

The first transmission, (1), is obviously an enumeration of the 32 different symbols which will be used in the communications; in transmission (2) is the clear implication that A represents the integer 1, B

[3] The Rosetta Stone is a piece of black basalt found in 1799 near the Rosetta mouth of the Nile, bearing a bilingual inscription (in Egyptian hieroglyphics, Egyptian demotic, and Greek) with which Jean François Champollion was able to solve the mystery of the Egyptian hieroglyphs.

(1) A. B. C. D. E. F. G. H. I. J. K. L. M. N. O. P. Q. R. S. T. U. V. W. X. Y. Z. •. &. $. ¢. ≠. @. A. B. C. D. E. F. G. H. I. J. K. L. M. N. O. P. Q. R. S. T. U. V. W. X. Y. Z. •. &. $. ¢. ≠. @.

(2) A A. B: A A A. C; A A A A. D; A A A A A. E; A A A A A A. F; A A A A A A A. G; A A A A A A A A. H; A A A A A A A A A. I; A A A A A A A A A A. J.

(3) A K A L B; A K A K A L C; A K A K A K A L D. A K A L B; B K A L C; C K A L D. B K C L E; E L B K C; F K D L J; J L D K F. E L K E; K E L E.

(4) C M A L B; D M A L C; G M E L B; E M G L M B.

(5) D K N L D; G K N L G; F M F L N; E M E L N.

(6) J L AN; J K A L AA; J K B L AB; AA K A L AB. J K J L BN; J K J K J L CN; IN K C L IC.

(7) B O C L F; D O B L H; E O B L AN; D O AN L DN.

(8) F P C L B; H P B L D; J P B L E; J P E L B.

(9) A P J L Q J; A P ANN L Q ANN; Q J. P J L Q ANN.

(10) Q J L R A; Q J O B L R B; A R E M A L R E L E O Q J. Q ANN L R NA; Q ANN O B L R NB.

(11) H L H; G S G. G S G. D K A L C K B; D K C S E K A; E K A S D K C.

(12) D T A; D T B; D T C; D L D; D U E; D U F; D U G. J T I; J U AA.

(13) FIRII V GN; ANNN K C V ANNN; AN P C V CNC.

(14) W E K A X L E K A; B W E K A X L W B O E X K W B O A X L B O F.

(15) C Y B L I; E Y B L BE; B Y E L CB; W D K A X Y B L EE.

(16) BE Z B L E; FD Z B L H; BC Z C L C; ABE Z C L E. W AI K F X Z B L E. BE Z B L M E; M ABE Z C L M E. BC Z B L NM E.

(17) D • L D O C O B O A L BD; E • L E O D O C O B O A L ABN; H • L DNCEN.

(18) & P D L A N Q C K Q E N Q G K Q I M. & V CRADAEI.

(19) $ L A K Q W A • X K Q W B • X K Q W C • X K Q W D • X K. $ V BRGAHGH.

(20) ¢ E K A ≠ L W E K A X; B ¢ E K A ≠ L B W E K A X; B ¢ E K W D K C X ≠ L B W E K G X. ¢ B ≠ ¢ D ≠ L W B X W D X L B O D.

(21) $ Y ¢ & O W K A X Z B ≠ K A L N.

(22) B K C L ¢ @NNA ≠; B K C L E. C O D L ¢ @NNA ≠; C O D L AB. D Y B L ¢ @NNA ≠; D Y B L AF.

(23) B K C L E; ¢ @NNB ≠. B K D L E; ¢ @NNC ≠. E Y B L BE; ¢ @NNB ≠. F Y B L C E; ¢ @NNC ≠. I T E; ¢ @NNB ≠. H U C; ¢ @NNC ≠. & V BRGAHBH; ¢ @NNC ≠.

(24) B L ¢ @NND ≠. C L ¢ @NND ≠. E. G. AA. AC. AG. ¢ @NND ≠. ANA ¢ @NND ≠.

(25) ¢ @NNE ≠ L B O & ¢ @A ≠; ¢ @NNF ≠ L & ¢ @A ≠ Y B. ¢ @A ≠ L ¢ @NNG ≠. ¢ @NNE ≠ L B & ¢ @NNG ≠; ¢ @NNF ≠ L & ¢ @NNG ≠ Y B.

(26) ¢ @NNH ≠ L D O Q C O & ¢ @NNG ≠ Y C.

(27) Q B K Q D K Q H K Q AF K Q CB V A; Q B K Q D K Q H K Q AF K Q CB K ¢ @NNI ≠ L A.

(28) C K ¢ @A ≠ L G; ¢ @A ≠ L D. I K ¢ @A ≠ L AB; ¢ @A ≠ L C. FD Z ¢ @A ≠ L H; ¢ @A ≠ L B.
¢ @A ≠ L A. R GG. M R GG. J P C. &. K. M. ¢ @NNI ≠.
¢ @B ≠ L A. R GG. M R GG. J P C. &. K. M. ¢ @NNI ≠.
¢ @C ≠ L A. R GG. M R GG. J P C. &. K. M. ¢ @NNI ≠.

(29) ¢ @NNE ≠ L B & ¢ @B ≠; ¢ @NNF ≠ L & ¢ @B ≠ Y B. ¢ @NAN ≠ L B ¢ @C ≠ K B ¢ @D ≠; ¢ @NAA ≠ L ¢ @C ≠ O ¢ @D ≠.

(30) ¢ @NNE ≠ L ¢ @NAB ≠; ¢ @NAD ≠; ¢ @NNF ≠ L ¢ @NAD ≠. ¢ @NAC ≠. ¢ @NAN ≠ L ¢ @NAB ≠. ¢ @NAE ≠; ¢ @NAA ≠ L ¢ @NAD ≠. ¢ @NAE ≠.

Fig. 3.

the integer 2, . . . , J the integer
there are introduced symbols for
in teaching us their mathema
addition, subtraction, multiplic
and the concept of zero; inequa
roots; and definitions of π and
new to the 31 symbols recovere
most beautiful concepts in pure
if they can teach us such a com
be staggered by what they wil
transmission. Beginning with
cluster concepts are introduce
transmission (30), we now are u
pure Venerean. Furthermore,
the code they are using on us
thousands upon thousands of
this is easily appreciated by an
the meaning of all 30 transmissi

Even right after this first mes
with that planet, we shall have
Fermat's Last Theorem, Gold
unsolved problems in mathem
not be difficult for "them" to d
nological superiority (first of a
able to call *us*!). If "they" b
structure constant," they are a
five for sure, suspect the sixth,
ratio, among others, of the spe
electron; it may take a century
And after we resolve our pres
appropriate to make discreet
harmony and peace with our fe
otherwise ingested by the su
fortune to contact us. But as
(and generations of his descend

[4] The solution may be found on p.

[5] With what he has learned fro
reader formulate these two questions d
compact form; the solutions appear on
classic unsolved problems in mathemati
values of x, y, and z can be found
is an integer greater than 2; Goldbach's
other mathematicians failed to make t
number greater than 2 can be expressed

N. O. P. Q. R. S. T. U. V. W. X. Y. Z.
F. G. H. I. J. K. L. M. N. O. P. Q. R.
@.

. E: A A A A A A. F: A A A A A A A. G:
I: A A A-A A A A A A A; J.

. L D. A K A L B: B K A L C: C K A L D.
K F. E L K E: K E L E.

L M B.

L N.

L AB. J K J L BN: J K J K J L CN:

N L DN.

L B.

L Q ANN.

E L E O Q J. Q ANN L R NA:

K C S E K A: E K A S D K C.

F: D U G. J T I: J U AA.

C.

E X K W B O A X L B O F.

A X Y B L BE.

Z C L E. W AI K F X Z B L E.
. NN E.

C C.B O A L ABN; H = L DNCEN.

V CRADAZI.

X K Q W D = X K. $ V ERGAHGH.

E K A X:
≠ ¢ D ≠ L W B X W D X L B O D.

@NNA ≠; C O D L AB.

≠. E Y B L BE; ¢ @NNB ≠. F Y B L C E:
@NNC ≠. & V ERGAHGH; ¢ @NNC ≠.
AC. AG. ¢ @NND ≠. ANA ¢ @NND ≠.
. ¢ @A ≠ Y B. ¢ @A ≠ L ¢ @NNG ≠.
¢ @NNG ≠ Y B.

≠ L A.

≠ L AB: ¢ @A ≠ L C.

¢ @NNI ≠.
¢ @NNI ≠.
¢ @NNI ≠.

@B ≠ Y B.
A ≠ L ¢ @C ≠ 0 ¢ @D ≠.

≠ L ¢ @NAD ≠. ¢ @NAC ≠.
≠ L ¢ @NAD ≠. ¢ @NAE ≠.

the integer 2, . . . , J the integer 10. In the first twenty transmissions there are introduced symbols for the introductory expository treatment in teaching us their mathematics. Among the items treated are: addition, subtraction, multiplication, and division; decimal notation and the concept of zero; inequalities and approximation; powers and roots; and definitions of π and e. Transmission (21) adds nothing new to the 31 symbols recovered thus far, but it does quote one of the most beautiful concepts in pure mathematics: they are telling us that, if they can teach us such a complex notion at this early stage, we will be staggered by what they will teach us by the 200th or the 2000th transmission. Beginning with transmission (22), words and word-cluster concepts are introduced, so that by the time we come to transmission (30), we now are understanding, in a manner of speaking, pure Venerean. Furthermore, we can now see how we could recover the code they are using on us, and which will obviously consist of thousands upon thousands of code groups with different meanings; this is easily appreciated by anyone who takes the trouble to fathom the meaning of all 30 transmissions in the foregoing example.[4]

Even right after this first message, if we are in direct communication with that planet, we shall have questions to put to "them": the proof of Fermat's Last Theorem, Goldbach's conjecture,[5] and many other unsolved problems in mathematics and the natural sciences. It will not be difficult for "them" to demonstrate their intellectual and technological superiority (first of all, don't forget it was *they* who were able to call *us*!). If "they" but know the *seventh* digit of the "fine structure constant," they are ages ahead of us (we know only the first five for sure, suspect the sixth). This number, 137.039 . . . , is the ratio, among others, of the speed of light to the speed of the hydrogen electron; it may take a century to calculate this constant to 9 digits. And after we resolve our pressing scientific questions, it might be appropriate to make discreet inquiries as to how we could live in harmony and peace with our fellow man—that is, if we aren't eaten or otherwise ingested by the superior civilization that had the good fortune to contact us. But as far as the cryptologist is concerned, he (and generations of his descendants who might experience the supreme

[4] The solution may be found on p. 109; but eschew the premature peek.

[5] With what he has learned from this example of space communication, let the reader formulate these two questions directly for transmission to "them," in a clear and compact form; the solutions appear on pg. 109. For the reader who is a little rusty on classic unsolved problems in mathematics, Fermat's Last Theorem states that no integral values of x, y, and z can be found to satisfy the equation $x^n + y^n = z^n$, if n is an integer greater than 2; Goldbach's "notorious" conjecture ("notorious" only because other mathematicians failed to make the conjecture themselves) states that every even number greater than 2 can be expressed as the sum of two primes.

thrill of their lives when we hear from "them") must keep a level head, not get excited, and be prepared to cope with problems the like of which he has never seen—out of this world, so to speak.

Electroni

BY JOHN

Top S

INTRODUCTION

Calligraphy, the art of produ
familiar term in the English lan
letters in its alphabet, English p
manual or mechanical. Slight
tions have little effect on legib
texts are easily obtained due to
All that is necessary is a unique
it is a typewriter key or a con
English requires only that a sig
letter. The Morse and Baudot
sult of such transmissions is eas

Generally speaking, most alp
graphic problems. However, s
graphic in nature. Such langu
sent some idea or thing. Altho
rules of formation, the number
often very large. For exampl
thousand unique ideographs. I
as a single element. The follow
lar symbols with very dissimilar r

Character

千 -------
米 -------
才 -------
木 -------
禾 -------

Therefore, for these languages,
tention in order to obtain intelli

Solution to Mr. Callimahos' Space Communication.

Symbols

A	1	I	9	Q	reciprocal	Y	power
B	2	J	10	R	decimal point	Z	root
C	3	K	+	S	≠	*	factorial
D	4	L	=	T	>	&	≠
E	5	M	−	U	<	$	e
F	6	N	ø	V	≈	¢	[
G	7	O	×	W	(	#	]
H	8	P	÷	X	)	@	code

Code values

001 question	007 radius	013 circle
002 true	008 volume or sphere	014 area
003 false	009 . . . (ellipsis)	015 rectangle
004 prime	010 perimeter of rect.	.
005 circum. of circle	011 area of rectangle	.
006 area of circle	012 perimeter	999

Code values 1, 2, 3 . . . 99 = x, y, z . . . (abstractions, unknowns, variables).

Fermat's Last Theorem:

¢@A # Y¢@D # K¢@B # Y¢@D # L¢@C # Y¢@D #. ¢@D ≠ L B;
¢@NNB #. ¢@D # L C, D, E, ¢@NNI #; ¢@NNB NNA #.

Goldbach's Conjecture:

B¢@A #. B¢@A # L¢@B #; ¢@B # T B. ¢@B # L¢@C # K¢@D #;
¢@C # L¢@NND #, ¢@D # L¢@NND #. ¢@NNB NNA #.

This page intentionally left blank.

* UFO Hypothesis and Survival Questions

This page intentionally left blank.

~~SECRET~~ 1968

DRAFT

U F O HYPOTHESIS AND SURVIVAL QUESTIONS

It is the purpose of this monograph to consider briefly some of the human survival implications suggested by the various principal hypothesis concerning the nature of the phenomena loosely categorized as U F O (1).

1. ALL UFO's ARE HOAXES: From the time when hoaxes were first noted in history, they were characterized by infrequency of occurrence and usually by a considerable restriction of their geographical extent. Rarely have men of science, while acting within their professional capacities, perpetrated hoaxes. The fact that UFO phenomenon have been witnessed all over the world from ancient times, and by considerable numbers of reputable scientists in recent times, indicates rather strongly that UFO's are not all hoaxes. (2) rather than diminishing, If anything, the modern trend is toward increased reports, from all sources. In one three month period in 1953 (June, July, and August) Air Force records show 35 sightings whose nature could not be determined (3). If UFO's, contrary to all

---

(1) All flying, sailing or maneuvering aerial objects whether glowing, pulsating, or of a constant metalic hue, whose shape is somewhat circular or cigarish.

(2) Anatomy of a Phenomenon, Jacques Vallee, Henry Regnery Co. Chicago, 1965. p 9-17. (Vallee has degrees in Astronomy and Physics and is currently consultant to NASA's MARS MAP STUDY.)

(3) United States Air Force Projects Grudge and Bluebook Reports 1-12 (1951-1953) National Investigation Committee on Aerial Phenomena. Washington June 1968. p 216.

(2) Visitors from Outer Space, Sputnik, (condensed from the almanac on Land and Sea) Vyacheslav Zaitsev, p 164-181.

DRAFT

~~SECRET~~ (1)

indications and expectations, are indeed hoaxes - hoaxes of a world-wide dimension - hoaxes of increasing frequency, then a human mental aberration of alarming proportions would appear to be developing. Such an aberration would seem to have serious implications for nations equipped with nuclear toys - and should require immediate and careful study by scientists.

2. ALL UFO's ARE HALLUCINATIONS: People, of course, do hallucinate. Although groups of people hallucinating is rare, it has been known to happen. Machines have their own form of hallucination; the radar, in particular, "sees" temperature inversions. But a considerable number of instances exist in which there are groups of people and a radar or radars seeing the same thing at the same time; sometimes a person and a guncamera confirm each other's testimony(4). On occasion, physical evidence of a circumstantial nature was reported to have been found to support witnessed sightings (5). A continuing high percentage of reports of unusual aerial objects are being reported by people in responsible positions in science, government, and industry.(6). The sum of such evidence seems to argue strongly against all UFO's being hallucinations. In spite of all the evidence to the contrary, if UFO's did turn out to be largely illusionary, the psychological implications for man would

on next page

(4) ibid., p 208, 192, 149, 146

(5) op. cit., Vallee, p. 70,71, 74.

(6) The Report on Unidentified Flying Objects, Edward J. Ruppelt, Doubleday, New York 1956. p 242. (Ruppelt was Chief of the Bluebook Study at one time.)

certainly bring into strong question his ability to distinguish reality from fantasy. The negative effect on man's ability to survive in an increasingly complex world would be considerable - making it imperative that such a growing impairment of the human capacity for rational judgment be subjected to immediate and thorough scientific study - so that the illness could be controlled before it reaches epidemic proportions (7). (For comments on mass hysteria and UFO's see source 8 below which contains a statement by Dr. Robert L. Hall, a social psychologist formerly with the AF Personnel and Training Research Center and the Program Director, Sociology and Psychology, National Science Foundation.)

3. ALL UFO's ARE NATURAL PHENOMENA: If this hypothesis is correct the capability of air warning systems to correctly diagnose an attack situation is open to serious question.

(→24)a. Many UFO's have been reported by trained military observers to behave like high speed, high performance, high altitude rockets or aircraft. The apparent solidity and craft-like shape of the objects have often been subject to radar confirmation (9). If such objects can appear to trained military men as rockets or air craft and if such objects

---

(7) op. cit., Ruppelt p 237 (Dutch Liner, Rome, Italy)

(8) Symposium on Unidentified Flying Objects (Second Session). July 29, 1968 nr 7 (House Committee on Science and Astronautics).

(9) op. cit., Project Grudge, pp.192, 149, 146.

SECRET DRAFT

should come over the Artic from the direction of Russia on the United States, they could trigger "false reports of missile attacks". (10)

b. Many responsible military officers have developed a mental "blind spot" to objects which appear to have the characteristics of UFO's. (10) Such an attitude is an open invitation to the enemy to build a replica of the phenomena in order to penetrate the "hole" in his adversaries' defenses - Was this the purpose of the lens shaped reentry vehicle tested by the U.S. Air Force in 1960 and recently featured in the Washington, D.C. Evening Star, dated 24 September 1968, page A4?

c. Sometimes the phenomena appear to defy radar detection and to cause massive electromagnetic interference. Surely it is very important to discover the nature of these objects or plasmas before any prospective enemy can use their properties to build a device or system to circumvent or jam our air and space detection systems - Any nation certainly could use a system or device to penetrate enemy defenses. (11)

4. SOME UFO's ARE SECRET EARTH PROJECTS: The above referenced U.S. Air Force reentry vehicle and an often publicized Canadian "saucer" project (9) leave little doubt as to the validity of this hypothesis. Undoubtedly, <u>all UFO's</u> should be carefully scrutinized to ferret-out such enemy (or "friendly") projects. Otherwise a nation faces the very strong next page

---

(10) <u>Flying Phenomena</u>, Sovietskaya Latviya, No. 287, 10 December 1967, p 3, Col 1-3, by R. Vitolniyek (Director of the Station for the Radio Observation of the Ionosphere and Artificial Earth Satellites).

(11) ibid., Project Grudge.

<u>UFO's and the Colorado Project</u>, Encyclopedia Brittanica Book of the Year 1968, p 123.

SECRET (11) DRAFT

possibility of being intimidated by a new secret "doomsday" weapon.

5. SOME UFO's ARE RELATED TO EXTRA-TERRESTRIAL INTELLIGENCE: According to some eminent scientists closely associated with the study of this phenomenon, this hypothesis cannot be disregarded. 12 (The well documented sightings over Wash. DC. in 1952 strongly support this view). 13 This hypothesis has a number of far-reaching human survival implications:

a. If "they" discover you, it is an old but hardly invalid rule of thumb, "they" are your technological superiors. Human history has shown us time and again the tragic results of a confrontation between a technologically superior civilization and a technologically inferior people. The "inferior" is usually subject to physical conquest.

b. Often in the past, a technologically superior people are also possessors of a more virile or aggressive culture. In a confrontation between two peoples of significantly different cultural levels, those having the inferior or less virile culture, most often suffer a tragic loss of identity and are usually absorbed by the other people.

c. Some peoples who were technologically and/or culturally inferior to other nations have survived - have maintained their identity - have equalized the differences between them and their adversaries. The Japanese people have given us an excellent example of the methods required to achieve such a survival:

(1) full and honest acceptance of the nature of the inferiorities separating you from the advantages of the other peoples,

(2) complete national solidarity in all positions taken in dealing with the other culture,

(3) highly controlled and limited intercourse with the other side - doing only those actions advantageous to the foreigner which you are absolutely forced to do by circumstances, next page

(4) a correct but friendly attitude toward the other people,

12. Professor James E MacDonald, Astronomer, Professor Allen J. Hynek, Astrophysicist Jacques Vallée Astronomer; Seymour Hess, Astronomer etc. that some of these objects are probably extra-terrestrial in origin is also supported by a 1958 report by Brazilian ~~[illegible]~~ scientific investigators.
13 Op. cit., The [illegible] (c) Washington Merry-Go-Round.

(5) A national eagerness to learn everything possible about the other culture - its technological and cultural strengths and weaknesses. This often involves sending selected groups and individuals to the other's country to become one of his kind, or even to help him in his wars against other adversaries.

(6) Adopting as many of the advantages of the opposing people as you can, and doing it as fast as possible - while still protecting your own identity by molding each new knowledge increment into your own cultural cast.

6. COMMENT: Although this paper has hardly exhausted the possible hypotheses related to the UFO phenomena, those mentioned above are the principal ones presently put forward. All of them have serious survival implications. The final answer to this mystery will probably include more than one of the above hypotheses.

Up until this time, the leisurely scientific approach has too often taken precedence in dealing with UFO questions. If you are walking along a forest path and someone yells, "rattler" your reaction would be immediate and defensive. You would not take time to speculate before you act. You would have to treat the alarm as if it were a real and immediate threat to your survival. Investigation would become an intensive emergency action to isolate the threat and to determine it's precise nature - It would be geared to developing adequate defensive measures in a minimum amount of time.

It would seem a little more of this survival attitude is called for in dealing with the UFO problem.

next page

DRAFT

~~SECRET~~ (6)

Observations of chimpanzees while in a captive environment have shown that the animals tend to become confused and disoriented. Since they do not usually have adult chimps to teach them how to be good apes, they are not even sure of their behavior. Often their actions are patterned after human behavior and would have virtually no survival value in the wild. Lacking the challenge of environmental adaptation, the bodies of the animals atrophy and become subject to many diseases - mostly unknown in their wild counterparts. Reactions to stimulis usually become less responsive and suitable. Sex often becomes a year-long preoccupation instead of a seasonal madness.

Do the captivity characteristics of modern civilization cause a similar lessening of man's adaptive capability, of his health, of his ability to recognize reality, of his ability to survive?

Perhaps the UFO question might even make man undertake studies which could enable him to construct a society which is most conducive to developing a completely <u>human</u> being, healthy in all aspects of mind and body - and most important able to recognize and adapt to real environmental situations.

[redacted], 51
x 2308
2143

This page intentionally left blank.

* UFO's and the Intelligence Community Blind Spot to Surprise or Deceptive Data

This page intentionally left blank.

~~SECRET~~ DRAFT

SUBJECT: UFO's and the Intelligence Community Blind Spot to Surprise or Deceptive Data

1. The implications of the UFO phenomena go far beyond the particular phenomena itself. The human incapacity to objectively process such data indicates a serious weakness in the intelligence community. This weakness ought to be remedied and quickly if the United States is to be able to respond swiftly and appropriately to surprise attack indicators. The reason that surprise attack is such a basic ingredient of military success is that it is able to rely on a most dependable human blind spot: The inability of most men to objectively process and evaluate highly unusual data and to react to the data in a meaningful way.

2. Scientific Findings: Dr. Jacques Vallee* famed communications science expert has studied thousands of cases where human beings have observed unusual phenomena. He has found that the human response to such observation is predictable and graphically depictable. Whether the person's psychological structure is being assaulted by the unusual and shocking brutality of a murder or the strangeness of a UFO sighting the effect is the same:

a. Initially as by a kind of psychological interia, the mind records fairly objectively what the eye is reporting.*

b. But when it has realized the strange nature of the phenomena it goes into shock. The mind likes to live in a comfortable world where it feels it knows what to expect, and that, is not too threatening either physically or psychologically. The unusual dispells the comfortable illusion the mind has created. This shock tears at the very mooring of the human psychological structure.*

c. To protect itself against such an intrusive and threatening reality the mind will begin to add imagination and interpretation to the incoming data to make it more acceptable. Since the mind is doing all this in haste some of the hurridly added details and suggestions tumble over one another and contradict one another in a bizzare fashion (as any police officer interrogating murder witnesses will tell you*) (See Chart A).

d. Once the mind has constructed a "safe" framework for the new information it may again peek out and collect some more objective data. If the data is still threatening it will again go into shock and the process starts all over again.*

e. If the data is at the highest strangeness level where it brings terror either:

(1) The mind will pass out and go into amnesia burying the events perhaps permanently in the unconscious.*

~~SECRET~~ DRAFT

DRAFT

[redacted]

(2) The personal psychological structure will collapse and the mind will reach down into its deepest place where "that which cannot be destroyed" is and it will abandon itself to this entity for survival protection. Encounter with this changeless indestructable entity is usually referred to as a religious experience. In the confusion and the shock, this experience is often attributed to the shocking event or object and that is why primative peoples worship such bizzare things as airplanes or cigarette lighters.

f. The degree of strangness of the phenomena dictates how many people the mind is willing and able to tell the event to. A mildly unusual or shocking event will be told to many people. A very shocking event of high strangness will be told to few people or practically none at all. Occasionally the event is so shockingly unusual that it isn't even reported to the person's conscious mind but is buried in the unconscious of the person where it is only accessible to hypnosis or careful level six communication sharing with another person. (See Chart B.)

[redacted]

4. Conclusions and Recommendations. It is apparent that we cannot allow such a human flaw to leave us blinded to unusual or surprising material. The example indicates that some people are less affected by strange phenomena than others, though still frightened by it, they remain capable of reporting it with a fair degree of objectivity. To solve this problem I recommend the following:

[redacted]

2

DRAFT

[redacted]

DRAFT

[redacted]

3

DRAFT

[redacted]

DRAFT

Appendix

Other Examples of Blindness to Surprise Material Causing Defeat

1. Because the Aztecs could not conceive of the Spanish as human enemies, they were destroyed by this sudden invasion of alien Gods who could not be defended against.

2. Because the Spanish could not believe that those crude Americans could build a modern effective naval force their outmoded ships were swiftly destroyed at Cuba and the Philippines in the Spanish-American War.

3. Because the French had such absolute faith in their maginot line they refused to believe that the radically new small swift blitzkreig army could outflank their line and shock them into a disorganized route in 1940.

4. Because we could not appreciate the power of the novel carrier strike force were surprised and defeated at Pearl Harbor.

5. Because Yammamoto was not aware that the most important ship in the modern fleet was the novel aircraft carrier, he protected his battleships by leaving them far behind the rest of the fleet - where they could afford no protection to the carriers, which without sufficient air cover were nearly defenseless against the American dive bombers at Midway.

6. Partly because the Germans believed that you could not mount an invasion without permanent port facilities they were firmly convinced Normandy invasion must really come at Pas de Calais. The result was that they refused to reinforce Normandy even after the beachhead was established. When they did finally move the armoured reserves to support the Normandy forces, Patton trapped them in the bag behind the Falise Gap. (It is interesting to note that the idea of portable port facilities was so novel and radical that the allies almost refused to accept it. It is no wonder that the Germans didn't take it into their calculations.

7. Because the Germans found the idea that any one would spend five million dollars to break their enigma machine to be inconceivable - too radical, they refused to accept the reality of the compromise of enigma even when they were presented with solid evidence to the contrary.

8. Because Americans refused to believe that fast erratic moving oval objects were anything but mental illusions, they were destroyed by fast erratic moving oval guided missiles from Russia.

DRAFT

[redacted]

[redacted]

DRAFT

9. Because Americans found it incomprehensible that a crude technology could effectively defend itself against a sophisticated weapons systems, many aircraft were lost to WWII model antiaircraft and small arms fire in Vietnam.

10. Because Americans considered that the "primitive" Vietnamese were not capable of fluent English and of using relatively sohpisticated techniques of deception against US forces - many artillery and air strikes were called onto friendly targets by enemy deception tactics which were often somewhat simple in procedure.

DRAFT

[redacted]

Chart A

The Pattern of Objective and Speculative
Material During Observations of Unusual Data

(Speculative Chart)

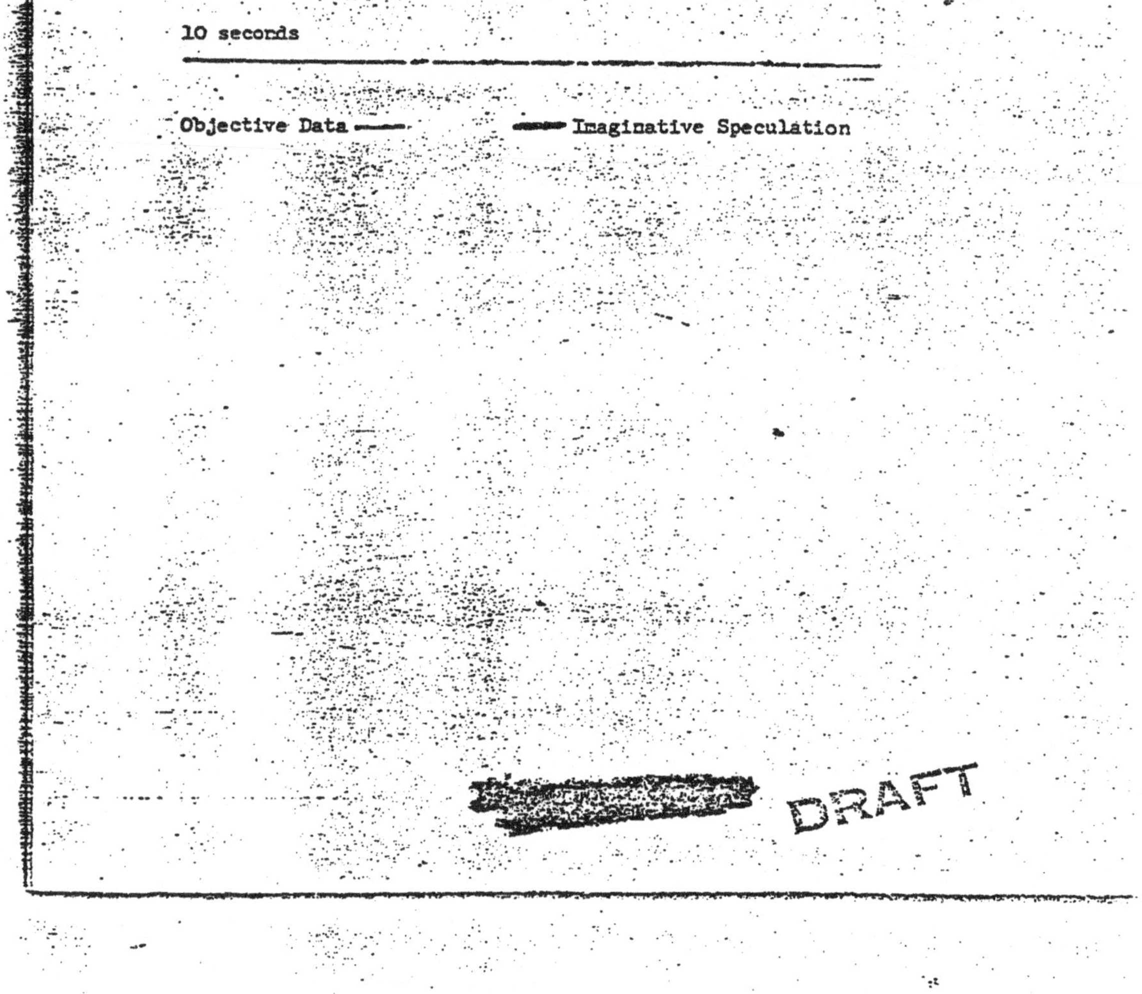

Chart B - Reporting Index

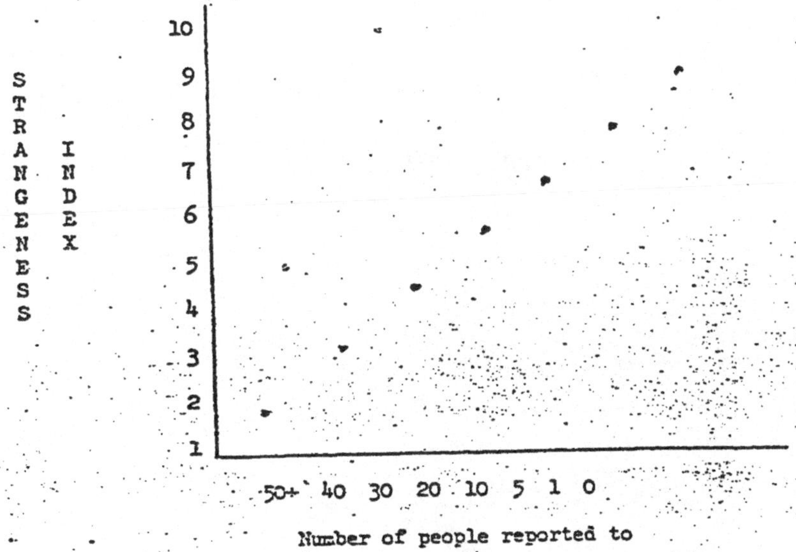

DRAFT.

This page intentionally left blank.

Memorandum for the Record

Subject:
Information Request Solicitation

This page intentionally left blank.

MEMORANDUM for the Record 29 August, 1978

From: [redacted] USN

To: D4

Subj: Information request solicitation

The enclosed chronology was drafted for the S group security officer and is forwarded for information.

[redacted]

The following is reported for the record. Background information is provided to maintain perspective.

<u>Monday, 10 July, 1978</u> Checked aboard Ft. Meade

<u>Friday, 21 July</u> I explained my association with UFOs to my supervisors and contacted [redacted] (former S1 staff) who inturn arranged an appointment with [redacted] of D4.

<u>Monday, 24 July</u> Discussed UFO status with D4 concluding that I understand my position, and that NSA prefers that I keep an extremely low profile.

<u>Wenesday, 26 July</u> Had a security brief by M5 where I explained my position on UFO matters and related that I had checked with D4 on this and that I had an understanding of my position relating UFOs, NSA and me. Discussed UFOs at length.

NOTE: Upon checking into a command I do not generally surface this problem till I am settled in but in this case I was due to attend a UFO Symposium in Dayton, Ohio on the 29th and 30th of July and thought it proper to notify NSA because of the remote possibility of the symposium getting some national media coverage and the possibility that my name or photograph might be associated with such coverage.

<u>Saturday and Sunday, 29-30 July</u> Attended the Mutual UFO Network 1978 UFO Symposium, Dayton, Ohio. On saturday a program was presented by a Mr. Stringfield and as an adjunct he introduced a Mr. Robert Barry. Barry discussed, and showed publicly, two letters he allegedly [redacted] My immediate opinion was that the letters were fraudulent. Sunday I was introduced to Mr. Stringfield and requested copies of the letters. I explained that I felt that the letters were frauds and that I thought I could possibly have them checked out.

<u>Thursday, 3 August</u> Contacted NSA [redacted] on the matter of the letters, who inturn referred me to [redacted] requested I provide them with copies of the letters, which I did, wherein they eventully

wrote to Mr. Barry explaining that the letters were fraudulent.

NOTE: During the period that I discussed the letters with [redacted] and some people in the UFO community, I had developed some suspicions, which I discussed with both groups, that:

1) [redacted] or,

2) [redacted] had some involvement because he claimed to know the signer of the letter that Barry recieved, which later on proved to be a non-existent entity, and he seemed, through secondhand information, to know too many answers to questions about the letters and [redacted]

[redacted]

<u>Sunday, 27 August</u> [redacted] Called me at approximately 2000. We talked for nearly one hour. He first wanted to clear up the matter of his working for NSA, a claim he has made publicly in several interviews. He said he had been talking to [redacted] and that [redacted] had said I had checked on his [redacted] employment at NSA. I explained to him that that was not the case but that in the process of letting the proper people know I was aboard, (D4), his name came up in relation to someone else requesting that information from NSA and that the fact was that NSA records did not reflect that he had every worked for them. He explained what his former association with NSA was. He had been, from 1963 to 1966, in ASA and at one time had been a shift supervisor at a crit-a-com relay [redacted] The discussion turned to general UFO topics and some general conversation [redacted] NSAs role, ie. they both provide a service and/or a product to many government agencies.

Then [redacted] ked me if it would be possible for me to do some looking around at NSA for UFO related information. He further explained that he did not mean covert looking but openly checking on what ever I could. I told him my standard answer, that the only thing I could do was to watch for information and to note its declassification date, then, if it was or would be declassified, I would either request the information myself or queue someone on the outside to request the information from the originator. He brought up the fact that "some things could be declassified simply by asking", suggesting that one can not ask for something that they did not know of its existence. I told him that in that case I would have to work that problem strickly from the inside untill such documents or information was declassified. I went on to explain to him that I work in COMSEC, which term he understood, and that I would have very little opportunity, if any, to ever find out if any such information existed. Further that in my experience with the Navy that in general I had seen very little information of interest to the UFO community and that when I had it was not in an official UFO context. He then explained that he understood that there was certain infromation that he, nor andone else in the civilian community, would ever see because of methods and/or sources, but, for example he knew that the National Command Center had sent UFO information to NSA as part of an address group.

All in all he was requesting me to watch out for UFO related information within NSA, which I tend to do on my own, but he was requesting that I pass on what I could. I, all in all, left him with the understanding that I was very limited in what I could do. I have, to some degree, checked on [redacted] character with some people who have worked with him more closely. There is some thought that he would be capable of being behind th[redacted] fraud and that he is apt to go to most any length to collect information or to bend facts to fit his needs.

3

I personall· ..have some fear that now that h has made contact with me, he may, either privately, or worse, publicly attempt to make it look as if I am an inside contact with NSA for him. Or, on the other hand, he may elude to having such a contact for years, then when he feels it appropriate or when cornered, hope to produce me as that contact.

I really do not know the man but I am fully confident that I can handle him in this situation. I have talked with him in a frank, clear manner that should have conveyed to him my position and I feel, without specific examples of his being dishonest, I should give him a chance of building a productive working relationship. As for his request for information, from what I know of the man I expected him to make such a forward approach upon meeting, though I did not anticipate him looking me up. I guess I should have expected that too.

Any further contact or requests for information will be reported.

[redacted]

4

This page intentionally left blank.

Unidentified Flying Objects and Air Force Project Blue Book

United States Air Force Fact Sheet 95-03

This page intentionally left blank.

FACT SHEET

USAF Fact Sheet 95-03

Unidentified Flying Objects and Air Force Project Blue Book

From 1947 to 1969, the Air Force investigated Unidentified Flying Objects under Project Blue Book. The project, headquartered at Wright-Patterson Air Force Base, Ohio, was terminated Dec. 17, 1969. Of a total of 12,618 sightings reported to Project Blue Book, 701 remained "unidentified."

The decision to discontinue UFO investigations was based on an evaluation of a report prepared by the University of Colorado entitled, *"Scientific Study of Unidentified Flying Objects;"* a review of the University of Colorado's report by the National Academy of Sciences; previous UFO studies and Air Force experience investigating UFO reports during the 1940s, '50s and '60s.

As a result of these investigations, studies and experience gained from investigating UFO reports since 1948, the conclusions of Project Blue Book were: (1) no UFO reported, investigated and evaluated by the Air Force was ever an indication of threat to our national security; (2) there was no evidence submitted to or discovered by the Air Force that sightings categorized as "unidentified" represented technological developments or principles beyond the range of modern scientific knowledge; and (3) there was no evidence indicating that sightings categorized as "unidentified" were extraterrestrial vehicles.

With the termination of Project Blue Book, the Air Force regulation establishing and controlling the program for investigating and analyzing UFOs was rescinded. Documentation regarding the former Blue Book investigation was permanently transferred to the Modern Military Branch, National Archives and Records Service, Eighth Street and Pennsylvania Avenue, N.W., Washington, D.C. 20408, and is available for public review and analysis.

Since the termination of Project Blue Book, nothing has occurred that would support a resumption of UFO investigations by the Air Force. Given the current environment of steadily decreasing defense budgets, it is unlikely the Air Force would become involved in such a costly project in the foreseeable future.

There are a number of universities and professional scientific organizations that have considered UFO phenomena during periodic meetings and seminars. A list of private organizations interested in aerial phenomena may be found in *Gale's Encyclopedia of Associations.* Interest in and timely review of UFO reports by private groups ensures that sound evidence is not overlooked by the scientific community. Persons wishing to report UFO sightings should be advised to contact local law enforcement agencies.

Point of Contact

News media requiring Project Blue Book files should contact the National Archives Public Affairs Office, (202)-501-5525. Public queries should be addressed to the Project Blue Book archivist at (202) 501-5385. For queries not related to Project Blue Book, contact the National Archives receptionist at (202) 501-5400.

(Current as of June 1995)

Report of
Air Force Research Regarding the
"Roswell Incident"

This page intentionally left blank.

REPORT OF AIR FORCE RESEARCH REGARDING THE

"ROSWELL INCIDENT"

July 1994

EXECUTIVE SUMMARY

The "Roswell Incident" refers to an event that supposedly happened in July, 1947, wherein the Army Air Forces (AAF) allegedly recovered remains of a crashed "flying disc" near Roswell, New Mexico. In February, 1994, the General Accounting Office (GAO), acting on the request of a New Mexico Congressman, initiated an audit to attempt to locate records of such an incident and to determine if records regarding it were properly handled. Although the GAO effort was to look at a number of government agencies, the apparent focus was on the Air Force. SAF/AAZ , as the Central Point of Contact for the GAO in this matter, initiated a systematic search of current Air Force offices as well as numerous archives and records centers that might help explain this matter. Research revealed that the "Roswell Incident" was not even considered a UFO event until the 1978-1980 time frame. Prior to that, the incident was dismissed because the AAF originally identified the debris recovered as being that of a weather balloon. Subsequently, various authors wrote a number of books claiming that, not only was debris from an alien spacecraft recovered, but also the bodies of the craft's alien occupants. These claims continue to evolve today and the Air Force is now routinely accused of engaging in a "cover-up" of this supposed event.

The research located no records at existing Air Force offices that indicated any "cover-up" by the USAF or any indication of such a recovery. Consequently, efforts were intensified by Air Force researchers at numerous locations where records for the period in question were stored. The records reviewed did not reveal any increase in operations, security, or any other activity in July, 1947, that indicated any such unusual event may have occurred. Records were located and thoroughly explored concerning a then-TOP SECRET balloon project, designed to attempt to monitor Soviet nuclear tests, known as Project Mogul. Additionally, several surviving project personnel were located and interviewed, as was the only surviving person who recovered debris from the original Roswell site in 1947, and the former officer who initially identified the wreckage as a balloon. Comparison of all information developed or obtained indicated that the material recovered near Roswell was consistent with a balloon device and most likely from one of the Mogul balloons that had not been previously recovered. Air Force research efforts did not disclose any records of the recovery of any "alien" bodies or extraterrestrial materials.

INTRODUCTION

Air Force involvement in the alleged UFO-related incident popularly known as the "Roswell Incident" began as the result of a January 14, 1994, Washington Post article (Atch 1) which announced Congressman Steven

Schiff's intent to initiate a General Accounting Office (GAO) effort to resolve this controversial matter. Having previously been involved in numerous Freedom of Information Act (FOIA) and Congressional requests on "unusual aircraft," to include Unidentified Flying Objects (UFOs), The Director, Security and Special Program Oversight, Office of the Secretary of the Air Force, (SAF/AAZ) believed the Air Force would become involved in any GAO effort involving this subject.

Thus, in late January, 1994, SAF/AAZ directed its research/declassification team, SAF/AAZD, to attempt to locate any official records relative to this matter. These initial research efforts focused on records at the Air Force Historical Research Agency (AFHRA), Maxwell AFB, AL, the Air Force Safety Agency (AFSA) at Kirtland AFB, NM, and the National Archives and Records Administration (NARA).

On February 15, 1994, the GAO officially notified Secretary of Defense William J. Perry that, it was initiating an audit of the Department of Defense (DoD) policies and procedures for acquiring, classifying, retaining, and disposing of official government documents dealing with weather balloon, aircraft, and similar crash incidents (Atch 2). This notification was sub- sequently passed to the Department of Defense Inspector General who in turn officially notified the Secretaries of the Services and other affected parties of the audit in a February 23, 1994, memo (Atch 3). This memorandum indicated that the "GAO is anxious to respond to Representative Schiff's request and to dispel any concerns that the DoD is being unresponsive." These were the first official US Government documents that indicated that the purpose of the GAO was to review "crash incidents involving weather balloons and unknown aircraft, such as UFOs and foreign aircraft, and (2) the facts involving the reported crash of an UFO in 1949 (sic, 1947) at Roswell, New Mexico ... (and an) alleged DoD cover-up."

An entrance meeting of potentially concerned parties was held in the offices of the DoD Inspector General on February 28, 1994. During this meeting it was learned that, while the audit officially would be reviewing the records of a number of DoD (and possibly other Executive Branch entities), the bulk of the effort would be focused on Air Force records and systems. The audit was officially given the GAO code 701034, and entitled "Records Management Procedures Dealing With Weather Balloon, Unknown Aircraft, and Similar Crash Incidents." Although this official title appeared rather broad, there was no misunderstanding that the real purpose was to attempt to locate records and/or information on the "Roswell Incident." This incident, explained later in more detail, generally dealt with the claim that in July of 1947, the US Army Air Forces (USAAF) recovered a flying saucer and /or its alien occupants which supposedly crashed near Roswell, New Mexico. When the USAAF ultimately became the United States Air Force (USAF) in September, 1947, the USAF inherited equipment, personnel, records, policies, and procedures from the AAF. In this particular case, the Air Force also inherited the allegation that it had "covered up" the "Roswell Incident" and has continued to do so for the next 47 years.

Within the Air Force, the Office of the Administrative Assistant to the Secretary of the Air Force (SAF/AA) is responsible both for information management procedures (SAF/AAI) and security policy and oversight (SAF/AAZ). Because of this organization, SAF/AA was the logical entity to assist the GAO in its audit and SAF/AAZ was officially named as the Central Point of Contact for this endeavor (Atch 4). Subsequently, the then Administrative Assistant, Mr. Robert J. McCormick, issued a tasking memorandum dated March 1, 1994 (Atch 5), to a number of current Air Staff and Secretariat offices that might possibly have records related to such an incident if, indeed, something had actually occurred. This search for records was purposely limited to Air Force records and systems since:

(a) The Air Force had no authority to compel other agencies to review their records;

(b) The Air Force would have no way to monitor the completeness of their efforts if they did; and

(c) the overall effort was the task and responsibility of the GAO--not the Air Force.

During the in-briefing process with GAO, it was learned that this audit was, indeed, generated at the specific request of Congressman Steven Schiff of New Mexico. Earlier, Congressman Schiff had written to the Department of Defense Legislative Liaison Office for information on the "Roswell Incident" and had been advised that it was part of the former UFO "Project Bluebook" that had previously been turned over to NARA by the Air Force. Congressman Schiff subsequently learned from NARA that, although they did, indeed, have the "Bluebook" materials, the "Roswell Incident" was not part of that report. Congressman Schiff, apparently perceiving that he had been "stonewalled" by the DoD, then generated the request for the aforementioned audit.

It is within this context that the following research and assistance efforts were conducted in support of the GAO. This report is intended to stand as the final official Air Force response regarding this matter.

THE "ROSWELL INCIDENT"--WHAT WAS ORIGINALLY REPORTED IN 1947

The modern preoccupation with what ultimately came to be called Unidentified Flying Objects (UFOs) actually began in June, 1947. Although some pro-UFO researchers argue that sightings of UFOs go back to Biblical times, most researchers will not dispute that anything in UFO history can compare with the phenomenon that began in 1947. What was later characterized as "the UFO Wave of 1947" began with 16 alleged sightings that occurred between May 17 and July 12, 1947, (although some researchers claim there were as many as 800 sightings during that period). Interestingly, the "Roswell Incident" was not considered one of these 1947 events until the 1978-1980 time frame. There is no dispute, however, that something happened near Roswell in July, 1947, since it was reported in a number of contemporary newspaper articles; the most famous of which were the July 8 and July 9 editions of the Roswell Daily Record.

The July 8 edition reported "RAAF Captures Flying Saucer On Ranch In Roswell Region," while the next day's edition reported, "Ramey Empties Roswell Saucer" and "Harassed Rancher Who Located 'Saucer' Sorry He Told About It."

The first story reported that the Intelligence Officer of the 509th Bomb Group, stationed at Roswell AAF, Major Jesse A. Marcel, had recovered a "flying disc" from the range lands of an unidentified rancher in the vicinity of Roswell and that the disc had been "flown to higher headquarters." That same story also reported that a Roswell couple claimed to have seen a large unidentified object fly by their home on July 2, 1947.

The July 9 edition of the paper noted that Brigadier General Roger Ramey, Commander of the Eighth Air Force at Forth Worth, Texas, stated that upon examination the debris recovered by Marcel was determined to be a weather balloon. The wreckage was described as a "..bundle of tinfoil, broken wood beams, and rubber remnants of a balloon...... The additional story of the "harassed rancher" identified him as W.W. Brazel of Lincoln County, New Mexico. He claimed that he and his son, Vemon, found the material on June 14, 1947, when they "came upon a large area of bright wreckage made up of rubber strips, tinfoil, a rather tough paper, and sticks." He picked up some of the debris on July 4 and ."..the next day he first heard about the flying discs and wondered if what he had found might have been the remnants of one of these." Brazel subsequently went to Roswell on July 7 and contacted the Sheriff, who apparently notified Major Marcel. Major Marcel and "a man in plain clothes" then accompanied Brazel home to pick up the rest of the pieces. The article further related that Brazel thought that the material:

> "..might have been as large as a table top. The balloon which held it up, if that is how it worked, must have been about 12 feet long, he felt, measuring the distance by the size of the room in which he sat. The rubber was smoky gray in color and scattered over an area about 200 yards in diameter. When the debris was gathered up the tinfoil, paper, tape, and sticks made a bundle about three feet long and 7 or 8 inches thick, while the rubber made a bundle about 18 or 20 inches long and about 8 inches thick. In all, he estimated, the entire lot would have weighed maybe five pounds. There was no sign of any metal in the area which night have been used for an engine and no sign of any propellers of any kind. Although at least one paper fin had been glued onto some of the tinfoil. There were no words to be found anywhere on the instrument although there were letters on some of the parts. Considerable scotch tape and some tape with flowers printed upon it had been used in the construction. No string or wire were to be found but there were some eyelets in the paper to indicate that some sort of attachment may have been used. Brazel said that he had previously found two weather balloons on the ranch, but that what he found this time did not in any way resemble either of these."

EVOLUTION OF THE EVENT FROM 1947 TO THE PRESENT

General Ramey's press conference and rancher Brazel's statement effectively ended this as a UFO-related matter until 1978, although some UFO researchers argue that there were several obtuse references to it in 1950's era literature. Roswell, for example, is not referred to in the official USAF investigation of UFOs reported in Project Bluebook or its predecessors, Project Sign and Project Grudge, which ran from 1948-1969 (which Congressman Schiff subsequently learned when he made his original inquiry).

In 1978, an article appeared in a tabloid newspaper, the National Inquirer, which reported the former intelligence officer, Marcel, claimed that he had recovered UFO debris near Roswell in 1947. Also in 1978, a UFO researcher, Stanton Friedman, met with Marcel and began investigating the claims that the material Marcel handled was from a crashed UFO. Similarly, two authors, William L. Moore and Charles Berlitz, also engaged in research which led them to publish a book, The Roswell Incident, in 1980. In this book they reported they interviewed a number of persons who claimed to have been present at Roswell in 1947 and professed to be either first or second hand witnesses to strange events that supposedly occurred. Since 1978-1980, other UFO researchers, most notably Donald Schmitt and Kevin Randle, claim to have located and interviewed even more persons with supposed knowledge of unusual happenings at Roswell. These included both civilian and former military persons.

Additionally, the Robert Stack-hosted television show "Unsolved Mysteries" devoted a large portion of one show to a "re-creation" of the supposed Roswell events. Numerous other television shows have done likewise, particularly during the last several years and a made-for-TV movie on the subject is due to be released this summer. The overall thrust of these articles, books and shows is that the "Roswell Incident" was actually the crash of a craft from another world, the US Government recovered it, and has been "covering up" this fact from the American public since 1947, using a combination of disinformation, ridicule, and threats of bodily harm, to do so. Generally, the US Air Force bears the brunt of these accusations.

From the rather benign description of the "event" and the recovery of some material as described in the original newspaper accounts, the "Roswell Incident" has since grown to mythical (if not mystical) proportions in the eyes and minds of some researchers, portions of the media and at least part of the American public. There are also now several major variations of the "Roswell story." For example, it was originally reported that there was only recovery of debris from one site. This has since grown from a minimal amount of debris recovered from a small area to airplane loads of debris from multiple huge "debris fields." Likewise, the relatively simple description of sticks, paper, tape and tinfoil has since grown to exotic metals with hieroglyphics and fiber optic-like materials. Most versions now claim that there were two crash sites where debris was recovered; and at the second site, alleged bodies of extraterrestrial aliens were supposedly retrieved. The number of these "alien bodies" recovered also varied. These claims are further complicated by the fact that UFO researchers are not in agreement among themselves as to exactly where these recovery sites were located or even the dates of the alleged crash(es).

Consistently, however, the AAF was accused of securing these sites, recovering all the material therefrom, keeping locals away, and returning the recovered wreckage (and bodies) to Roswell under extremely tight security for further processing and later exploitation.

Once back at Roswell AAF, it is generally alleged that special measures were taken to notify higher headquarters and arrangements made to have recovered materials shipped to other locations for analysis. These locations include Ft. Worth, Texas, the home of the Eighth Air Force Headquarters; possibly Sandia Base (now Kirtland AFB), New Mexico; possibly Andrews AAF, Maryland, and always to Wright Field, now known as Wright Patterson AFB, Ohio. The latter location was the home of "T-2" which later became known as the Air Technical Intelligence Center (ATIC) and the Air Materiel Command (AMC), and would, in fact, be a logical location to study unknown materials from whatever origin. Most of the Roswell stories that contain the recovery of alien bodies also show them being shipped to Wright Field. Once the material and bodies were dispersed for further analysis and/or exploitation, the government in general, and the Army Air Forces in particular, then engaged in covering up all information relating to the alleged crash and recovery, including the use, of security oaths to military persons and the use of coercion (including alleged death threats) to others. This, as theorized by some UFO researchers, has allowed the government to keep the fact that there is intelligent extra-terrestrial life from the American public for 47 years. It also supposedly allowed the US Government to exploit recovered extraterrestrial materials by reverse engineering them, ultimately providing such things as fiber optic and stealth technology. The "death threats," oaths, and other forms of coercion alleged to have been meted out by the Army Air Forces personnel to keep people from talking have apparently not been very effective, as several hundred people are claimed to have come forward (without harm) with some knowledge of the "Roswell Incident" during interviews with non-government researchers and the media.

Adding some measure of credibility to the claims that have arisen since 1978 is the apparent depth of research of some of the authors and the extent of their efforts. Their claims are lessened somewhat, however, by the fact that almost all their information came from verbal reports many years after the alleged incident occurred. Many of the persons interviewed were, in fact, stationed at, or lived near Roswell during the time in question, and a number of them claim military service. Most, however, related their stories in their older years, well after the fact. In other cases, the information provided is second or thirdhand, having been passed through a friend or relative after the principal had died. What is uniquely lacking in the entire exploration and exploitation of the "Roswell Incident" is official positive documentary or physical evidence of any kind that supports the claims of those who allege that something unusual happened. Conversely, there has never been any previous documentary evidence produced by those who would debunk the incident to show that something did not happen; although logic dictates that bureaucracies do not spend time documenting non-events.

SEARCH STRATEGY AND METHODOLOGY

To insure senior Air Force leadership that there were no hidden or overlooked files that might relate to the "Roswell Incident;" and to provide the GAO with the best and most complete information available, SAF/AAZ constructed a strategy based on direct tasking from the Office of the Secretary, to elicit information from those functional offices and organizations where such information might logically be contained. This included directing searches at current offices where special or unusual projects might be carried out, as well as historical organizations, archives, and records centers over which the Air Force exerted some degree of control. Researchers did not, however, go to the US Army to review historical records in areas such as missile launches from White Sands, or to the Department of Energy to determine if its forerunner, the Atomic Energy Commission, had any records of nuclear-related incidents that might have occurred at or near Roswell in 1947. To do so would have encroached on GAO's charter in this matter. What Air Force researchers did do, however, was to search for records still under Air Force control pertaining to these subject areas.

In order to determine parameters for the most productive search of records, a review was first conducted of the major works regarding the "Roswell Incident" available in the popular literature. These works included: The Roswell Incident, (1980) by William Moore and Charles Berlitz; "Crashed Saucers: Evidence in Search of Proof," (1985) by Moore; The UFO Crash at Roswell, (I 99 1) by Kevin Randle and Donald Schmitt; The Truth About the UFO Crash at Roswell, (1994) also by Randle and Schmitt; The Roswell Report: A Historical Perspective, (1991), George M. Eberhart, Editor; "The Roswell Events," (1993) compiled by Fred Whiting- Crash at Corona (1992) by Stanton T. Friedman and Don Berliner, as well as numerous other articles written by a combination of the above and other researchers. Collectively, the above represent the "pro" UFO writers who allege that the government is engaged in a conspiracy. There are no specific books written entirely on the theme that nothing happened at Roswell. However, Curtis Peebles in Watch the Skies! (1994) discussed the development of the UFO story and growth of subsequent claims as a phenomenon. There has also been serious research as well as a number of detailed articles written by so-called "debunkers" of Roswell and other incidents, most notably Philip J. Klass who writes The Skeptical Inquirer newsletter, and Robert Todd, a private researcher. The concerns and claims of all the above authors and others were considered in conducting the USAF records search.

It was also decided, particularly after a review of the above popular literature, that no specific attempt would be made to try to refute, point by point, the numerous claims made in the various publications. Many of these claims appear to be hearsay, undocumented, taken out of context, self-serving, or otherwise dubious. Additionally, many of the above authors are not even in agreement over various claims. Most notable of the confusing and now ever-changing claims is the controversy over the date(s) of the alleged incident, the exact location(s) of the purported debris and the extent of the wreckage. Such discrepancies in claims made the search much more difficult by greatly expanding the volume of records that had to be searched.

An example of trying to deal with questionable claims is illustrated by the following example: One of the popular books mentioned that was reviewed claimed that the writers had submitted the names and serial numbers of "over two dozen" personnel stationed at Roswell in July, 1947, to the Veterans Administration and the Defense Department to confirm their military service. They then listed eleven of these persons by name and asked the question "Why does neither the Defense Department nor the Veteran's Administration have records of any of these men when we can document that each served at Roswell Army Air Field." That claim sounded serious so SAF/AAZD was tasked to check these eleven names in the Personnel Records Center in St. Louis. Using only the names (since the authors did not list the serial numbers) the researcher quickly found records readily identifiable with eight of these persons. The other three had such common names that there could have been multiple possibilities. Interestingly, one of the listed "missing" persons had a casualty report in his records reflecting that he died in 195 1, while the writers claimed to have interviewed him (or a person of the exact same name) in 1990.

While the historical document search was in progress, it was decided to attempt to locate and interview several persons identified as still living who could possibly answer questions generated by the research. This had never been officially done before, although most of the persons contacted reported that they had also been contacted in the past by some of the listed authors or other private researchers. In order to counter possible future arguments that the persons interviewed were still "covering up" material because of prior security oaths, the interviewees were provided with authorization from either the Secretary of the Air Force or the Senior Security Official of the Air Force that would officially allow discussion of classified information, if applicable, or free them from any prior restriction in discussing the matter, if such existed. Again, the focus was on interviewing persons that could address specific issues, raised by research and no consideration was given to try and locate every alleged witness claimed to have been contacted by the various authors. For example, one of the interviewees thought vital to obtain an official signed,' sworn statement from was Sheridan Cavitt, Lt Col, USAF (Retired) who is the last living member of the three persons universally acknowledged to have recovered material from the Foster Ranch. Others were also interviewed as information developed (discussed in detail later). Additionally, in some cases survivors of deceased persons were also contacted in an attempt to locate various records thought to have been in the custody of the deceased.

Even though Air Force research originally started in January, 1994, the first official Air Force-wide tasking was directed by the March 1, 1994, memorandum from

SAF/AA, (Atch 5) and was addressed to those current Air Staff elements that would be the likely repository for any records, particularly if there was anything of an extraordinary nature involved. This meant that the search was not limited to unclassified materials, but also would include records of the highest classification and compartmentation.

The specific Air Staff/Secretariat offices queried included the following:

(a) SAF/AAI, Directorate of Information Management

(b) SAF/AQL, Directorate of Electronics and Special Programs

(c) AF/SE, Air Force Safety

(d) AF/HO, Air Force Historian

(e) AF/IN, Air Force Intelligence (including Air Force Intelligence Agency--AFIA, and the National Air Intelligence Center, NAIC)

(f) AF/XOW, Directorate of Weather

(g) (added later) The Air Force Office of Special Investigations (AFOSI)

In addition to the above Air Staff and Secretariat offices, SAF/AAZ also reviewed appropriate classified records for any tie-in to this matter. With regards to highly classified records, it should be noted that any programs that employ enhanced security measures or controls are known as a Special Access Programs (SAPs). The authority for such programs comes from Executive Order 12356 and flows from the Department of Defense to the Services via DoD Directive 5205.7. These programs are implemented in the Air Force by Policy Directive 16-7, and Air Force Instruction 16-701. These directives contain detailed requirements for controlling and reporting, in a very strict manner, all SAPS. This includes a report from the Secretary of the Air Force to the Secretary of Defense (and ultimately to Congress) on all SAPs submitted for approval, and a certification that there are no "SAP-like" programs being operated. These reporting requirements are stipulated in public law.

It followed then, that if the Air Force had recovered some type of extraterrestrial spacecraft and/or bodies and was exploiting this for scientific and technology purposes, then such a program would be operated as a SAP. SAF/AAZ, the Central Office for all Air Force SAPs, has knowledge of, and security oversight over, all SAPs. SAF/AAZ categorically stated that no such Special Access Program(s) exists that pertain to extraterrestrial spacecraft/aliens.

Likewise, the Secretary of the Air Force and the Chief of Staff, who head the Special Program Oversight Committee which oversees all sensitive programs in the Air Force, had no knowledge of the existence of any such program involving, or relating to the events at Roswell or the alleged technology that supposedly resulted therefrom. Besides the obvious irregularity and illegality of keeping such information from the most senior Air Force, officials, it would also be illogical, since these officials are responsible for obtaining funding for operations, research, development, and security. Without funding such a program, operation, or organization could not exist. Even to keep such a fact "cover-up" in some sort of passive "caretaker status" would involve money. More importantly, it would involve people and create paperwork.

The aforementioned March 1, 1994, SAF/AA tasking generated negative responses (Atch 6-12) from all recipients; i.e. all offices reported that they had no information that would explain the incident. Consequently, these negative responses led to an increase in the already on-going historical research at records centers and archives.

The extensive archival and records center search was systematically carried out at by the SAF/AAZD Declassification Review Team. This team is composed entirely of Air Force Reserve personnel who have extensive training and experience in large scale review of records. (Previous efforts include the Southeast Asia Declassification Review, declassification of POW/MIA records, and the review of the Gulf War Air Power Survey records). The team members all had the requisite security clearances for classified information and had the authority of the Secretary of the Air Force to declassify any classified record they found that might be related to Roswell. SAF/AAZD conducted reviews at a number of locations, including: the National Archives in Washington, DC; the National Personnel Records Center, St. Louis, MO; the National Archives, Suitland, MD, the National Records Center, Suitland, MD; Naval Research Laboratory, Washington, DC; Federal Records Center, Ft Worth, TX; the INSCOM Archives, Ft. Meade, MD; National Air and Space Museum, Washington, DC; Air Force Historical Research Agency, Maxwell AFB, AL; Center for Air Force History, Bolling AFB, DC; Phillips Laboratory, Hanscom AFB, MA, and Kirtland AFB, NM; Rome Laboratory, Griffiss AFB, NY; and the Library of Congress, Washington, DC.

A listing of the specific record areas searched is appended as Atch 13. The areas included all those subject areas logically believed to possibly contain any reference to activities at Roswell AAF during the period of time in question. It is anticipated that detractors from this effort will complain that "they did not search record group x , box y, or reel z, etc.; that's where the real records are!" Such complaints are unavoidable and there is no possible way that the millions of records under Air Force control could be searched page by page. The team endeavored to make logical searches in those places where records would likely be found. They were assisted in this task by archivists, historians, and records management specialists, including experienced persons who have continually worked in Army and Air Force records systems since 1943. The team also searched some record areas that were recommended by serious private researchers such as Robert Todd, who had independently obtained almost encyclopedic knowledge of the complexities of Air Force records systems, particularly as related to this subject area.

Not surprisingly, the research team found the usual number of problems in many of the records centers (particularly St. Louis) with misfiling, lost or misplaced documents, mismarking of documents, or the breaking up of record groups over the years and refiling in different systems. This included, for example, a small amount of missing "decimal files" from the 509th Bomb Group at Roswell that covered the years 1945-1949, that were marked on the index as "destroyed." The researchers noted that there was no pattern to any anomalies found and that most discrepancies were minor and consistent with what they had found in the past on similar projects.

WHAT THE ROSWELL INCIDENT WAS NOT

Before discussing specific positive results that these efforts revealed, it is first appropriate to discuss those things, as indicated by information available to the Air Force, that the "Roswell Incident" was not:

An Airplane Crash

Of all the things that are documented and tracked within the Air Force, among the most detailed and scrupulous are airplane crashes. In fact, records of air crashes go back to the first years of military flight. Safety records and reports are available for all crashes that involved serious damage, injury, death, or a combination of these factors. These records also include incidents involving experimental or classified aircraft. USAF records showed that between June 24, 1947, and July 28, 1947, there were five crashes in New Mexico alone, involving A-26C, P-5 IN, C-82A, P-80A and PQ-14B aircraft; however, none of these were on the date(s) in question nor in the area(s) in question.

One of the additional areas specifically set forth by GAO in its efforts was to deal with how the Air Force (and others) specifically documented ."..weather balloon...and other crash incidents." In this area, the search efforts revealed that there are no air safety records pertaining to weather balloon crashes (all weather balloons "crash" sooner or later); however, there are provisions for generating reports of "crashes" as ground safety incidents in the unlikely chance that a balloon injures someone or causes damage. However, such records are only maintained for five years.

A Missile Crash

A crashed or errant missile, usually described as a captured German V-2 or one of its variants, is sometimes set forth as a possible explanation for the debris recovered near Roswell. Since much of this testing done at nearby White Sands was secret at the time, it would be logical to assume that the government would handle any missile mishap under tight security, particularly if the mishap occurred on private land. From the records reviewed by the Air Force, however, there was nothing located to suggest that this was the case. Although the bulk of remaining testing records are under the control of the US Army, the subject has also been very well documented over the years within Air Force records. There would be no reason to keep such information classified today. The USAF found no indicators or even hints that a missile was involved in this matter.

A Nuclear Accident

One of the areas considered was that whatever happened near Roswell may have involved nuclear weapons. This was a logical area of concern since the 509th Bomb Group was the only military unit in the world at the time that had access to nuclear weapons. Again, reviews of available records gave no indication that this was the case. A number of records still classified TOP SECRET and SECRET-RESTRICTED DATA having to do with nuclear weapons were located in the Federal Records Center in St. Louis, MO . These records, which pertained to the 509th, had nothing to do with any activities that could have been misinterpreted as the "Roswell Incident." Also, any records of a nuclear related incident would have been inherited by the Department of Energy (DOE), and, had one occurred, it is likely DOE would have publicly reported it as part of its recent declassification and public release efforts. There were no ancillary records in Air Force files to indicate the potential existence of such records within DOE channels, however.

An Extraterrestrial Craft

The Air Force research found absolutely no indication that what happened near Roswell in 1947, involved any type of extraterrestrial spacecraft. This, of course, is the crux of this entire matter. "Pro-UFO" persons who obtain a copy of this report, at this point, most probably begin the "cover-up is still on" claims. Nevertheless, the research indicated absolutely no evidence of any kind that a spaceship crashed near Roswell or that any alien occupants were recovered therefrom, in some secret military operation or otherwise. This does not mean, however, that the early Air Force was not concerned about UFOs. However, in the early days, "UFO" meant Unidentified Flying Object, which literally translated as some object in the air that was not readily identifiable. It did not mean, as the term has evolved in today's language, to equate to alien spaceships. Records from the period reviewed by Air Force researchers as well as those cited by the authors mentioned before, do indicate that the USAF was seriously concerned about the inability to adequately identify unknown flying objects reported in American airspace. All the records, however, indicated that the focus of concern was not on aliens, hostile or otherwise, but on the Soviet Union. Many documents from that period speak to the possibility of developmental secret Soviet aircraft overflying US airspace. This, of course, was of major concern to the fledgling USAF, whose job it was to protect these same skies.

The research revealed only one official AAF document that indicated that there was any activity of any type that pertained to UFOs and Roswell in July, 1947. This was a small section of the July Historical Report for the 509th Bomb Group and Roswell AAF that stated: "The Office of Public Information was quite busy during the month answering inquiries on the 'flying disc,' which was reported to be in possession of the 509th Bomb Group. The object turned out to be a radar tracking balloon" (included with Atch I 1). Additionally, this history showed that the 509th Commander, Colonel

Blanchard, went on leave on July 8, 1947, which would be a somewhat unusual maneuver for a person involved in the supposed first ever recovery of extraterrestrial materials. (Detractors claim Blanchard did this as a ploy to elude the press and go to the scene to direct the recovery operations). The history and the morning reports also showed that the subsequent activities at Roswell during the month were mostly mundane and not indicative of any unusual high level activity, expenditure of manpower, resources or security.

Likewise, the researchers found no indication of heightened activity anywhere else in the military hierarchy in the July, 1947, message traffic or orders (to include classified traffic). There were no indications and warnings, notice of alerts, or a higher tempo of operational activity reported that would be logically generated if an alien craft, whose intentions were unknown, entered US territory. To believe that such operational and high-level security activity could be conducted solely by relying on unsecured telecommunications or personal contact without creating any records of such activity certainly stretches the imagination of those who have served in the military who know that paperwork of some kind is necessary to accomplish even emergency, highly classified, or sensitive tasks.

An example of activity sometimes cited by pro-UFO writers to illustrate the point that something unusual was going on was the travel of Lt. General Nathan Twining, Commander of the Air Materiel Command, to New Mexico in July, 1947. Actually, records were located indicating that Twining went to the Bomb Commanders' Course on July 8, along with a number of other general officers, and requested orders to do so a month before, on June 5, 1947 (Atch 14).

Similarly, it has also been alleged that General Hoyt Vandenberg, Deputy Chief of Staff at the time, had been involved directing activity regarding events at Roswell. Activity reports (Atch 15), located in General Vandenberg's personal papers stored in the Library of Congress, did indicate that on July 7, he was busy with a "flying disc" incident; however this particular incident involved Ellington Field, Texas and the Spokane (Washington) Depot. After much discussion and information gathering on this incident, it was learned to be a hoax. There is no similar mention of his personal interest or involvement in Roswell events except in the newspapers.

The above are but two small examples that indicate that if some event happened that was one of the "watershed happenings" in human history, the US military certainly reacted in an unconcerned and cavalier manner. In an actual case, the military would have had to order thousands of soldiers and airman, not only at Roswell but throughout the US, to act nonchalantly, pretend to conduct and report business as usual, and generate absolutely no paperwork of a suspicious nature, while simultaneously anticipating that twenty years or more into the future people would have available a comprehensive Freedom of Information Act that would give them great leeway to review and explore government documents. The records indicate that none of this happened (or if it did, it was controlled by a security system so efficient and tight that no one, US or otherwise, has been able to duplicate it since. If such a system had been in effect at the time, it would have also been used to protect our atomic secrets from the Soviets, which history has showed obviously was not the case). The records reviewed confirmed that no such sophisticated and efficient security system existed.

WHAT THE "ROSWELL INCIDENT" WAS

As previously discussed, what was originally reported to have been recovered was a balloon of some sort, usually described as a "weather balloon," although the majority of the wreckage that was ultimately displayed by General Ramey and Major Marcel in the famous photos (Atch 16) in Ft. Worth, was that of a radar target normally suspended from balloons. This radar target, discussed in more detail later, was certainly consistent with the description of July 9 newspaper article which discussed "tinfoil, paper, tape, and sticks." Additionally, the description of the "flying disc" was consistent with a document routinely used by most pro-UFO writers to indicate a conspiracy in progress--the telegram from the Dallas FBI office of July 8, 1947. This document quoted in part states: ."..The disc is hexagonal in shape and was suspended from a balloon by a cable, which balloon was approximately twenty feet in diameter. ...the object found resembles a high altitude weather balloon with a radar reflector. ...disc and balloon being transported..."

Similarly, while conducting the popular literature review, one of the documents reviewed was a paper entitled "The Roswell Events" edited by Fred Whiting, and sponsored by the Fund for UFO Research (FUFOR). Although it was not the original intention to comment on what commercial authors interpreted or claimed that other persons supposedly said, this particular document was different because it contained actual copies of apparently authentic sworn affidavits received from a number of persons who claimed to have some knowledge of the Roswell event. Although many of the persons who provided these affidavits to the FUFOR researchers also expressed opinions that they thought there was something extraterrestrial about this incident, a number of them actually described materials that sounded suspiciously like wreckage from balloons. These included the following:

Jesse A. Marcel, NM (son of the late Major Jesse Marcel; 11 years old at the time of the incident). Affidavit dated May 6, 1991. " ... There were three categories of debris: a thick, foil like metallic gray substance; a brittle, brownish-black plastic-like material, like Bakelite; and there were fragments of what appeared to be I-beams. On the inner surface of the I-beam, there appeared to be a type of writing. This writing was a purple-violet hue, and it had an embossed appearance. The figures were composed of curved, geometric shapes. It had no resemblance to Russian, Japanese or any other foreign language. It resembled hieroglyphics, but it had no animal-like characters ......

Loretta Proctor (former neighbor of rancher W.W. Brazel). Affidavit dated May 5, 199 1. ."..Brazel came to my ranch and showed my husband and me a piece of material he said came from a large pile of debris on the property he managed. The piece he brought was brown in color, similar to plastic...'Mac' said the other material on the property looked like aluminum foil. It was very flexible and wouldn't crush or bum. There was also something he described as tape which had printing on it. The color of the printing was a kind of purple..."

Bessie Brazel Schreiber (daughter of W.W. Brazel; 14 years old at the time of the incident). Affidavit dated September 22, 1993. ."..The debris looked like pieces of a large balloon which had burst. The pieces were small, the largest I remember measuring about the same as the diameter of a basketball. Most of it was a kind of double-sided material, foil-like on one side and rubber-like on the other. Both sides were grayish silver in color, the foil more silvery than the rubber. Sticks, like kite sticks, were attached to some of the pieces with a whitish tape. The tape was about two or three inches wide and had flowerlike designs on it. The 'flowers' were faint, a variety of pastel colors, and reminded me of Japanese paintings in which the flowers are not all connected. I do not recall any other types of material or markings, nor do I remember seeing gouges in the ground or any other signs that anything may have hit the ground hard. The foil-rubber material could not be tom like ordinary aluminum foil can be tom..."

Sally Strickland Tadolini (neighbor of WW Brazel; nine years old in 1947). Affidavit dated September 27, 1993. ".. What Bill showed us was a piece of what I still think as fabric. It was something like aluminum foil, something like satin, something like welltanned leather in its toughness, yet was not precisely like any one of those materials. ...It was about the thickness of very fine kidskin glove leather and a dull metallic grayish silver, one side slightly darker than the other. I do not remember it having any design or embossing on it..."

Robert R. Porter (B-29 flight Engineer stationed at Roswell in 1947). Affidavit dated June 7, 1991 " On this occasion, I was a member of the crew which flew parts of what we were told was a flying saucer to Fort Worth. The people on board included ... and Maj Jesse Marcel. Capt. William E. Anderson said it was from a flying saucer. After we arrived, the material was transferred to a B-25. I was told they were going to Wright Field in Dayton, Ohio. I was involved in loading the B-29 with the material, which was wrapped in packages with wrapping paper. One of the pieces was triangle-shaped, about 2 1/2 feet across the bottom. The rest were in small packages, about the size of a shoe box. The brown paper was held with tape. The material was extremely lightweight. When I picked it up, it was just like picking up an empty package. We loaded the triangle shaped package and three shoe box-sized packages into the plane. All of the packages could have fit into the trunk of a car. ...When we came back from lunch, they told us they had transferred the material to a B-25. They told us the material was a weather balloon, but I'm certain it wasn't a weather balloon..."

In addition to those persons above still living who claim to have seen or examined the original material found on the Brazel Ranch, there is one additional person who was universally acknowledged to have been involved in its recovery, Sheridan Cavitt, Lt Col, USAF, (Ret) . Cavitt is credited in all claims of having accompanied Major Marcel to the ranch to recover the debris, sometimes along with his Counter Intelligence Corps (CIC) subordinate, William Rickett, who, like Marcel, is deceased. Although there does not appear to be much dispute that Cavitt was involved in the material recovery, other claims about him prevail in the popular literature. He is sometimes portrayed as a closed-mouth (or sometimes even sinister) conspirator who was one of the early individuals who kept the "secret of Roswell" from getting out. Other things about him have been alleged, including the claim that he wrote a report of the incident at the time that has never surfaced.

Since Lt Col Cavitt, who had first-hand knowledge, was still alive, a decision was made to interview him and get a signed sworn statement from him about his version of the events. Prior to the interview, the Secretary of the Air Force provided him with a written authorization and waiver to discuss classified information with the interviewer and release him from any security oath he may have taken. Subsequently, Cavitt was interviewed on May 24, 1994, at his home. Cavitt provided a signed, sworn statement (Atch 17) of his recollections in this matter. He also consented to having the interview tape-recorded. A transcript of that recording is at Atch 18. In this interview, Cavitt related that he had been contacted on numerous occasions by UFO researchers and had willingly talked with many of them; however, he felt that he had oftentimes been misrepresented or had his comments taken out of context so that their true meaning was changed. He stated unequivocally, however, that the material he recovered consisted of a reflective sort of material like aluminum foil, and some thin, bamboo-like sticks. He thought at the time, and continued to do so today, that what he found was a weather balloon and has told other private researchers that. He also remembered finding a small "black box" type of instrument, which he thought at the time was probably a radiosonde. Lt Col Cavitt also reviewed the famous Ramey/Marcel photographs (Atch 16) of the wreckage taken to Ft. Worth (often claimed by LITO researchers to have been switched and the remnants of a balloon substituted for it) and he identified the materials depicted in those photos as consistent with the materials that he recovered from the ranch. Lt Col Cavitt also stated that he had never taken any oath or signed any agreement not to talk about this incident and had never been threatened by anyone in the government because of it. He did not even know the incident" was claimed to be anything unusual until he was interviewed in the early 1980's.

Similarly, Irving Newton, Major, USAF, (Ret) was located and interviewed. Newton was a weather officer assigned to Fort Worth, who was on duty when the Roswell debris was sent there in July, 1947. He was told that he was to report to General Ramey's office to view the material. In a signed, sworn statement (Atch 30) Newton related that ."..I walked into the General's office where this supposed flying saucer was lying all over the floor. As soon as I saw it, I giggled and asked if that was the flying saucer ... I told them that this was a balloon and a RAWIN target..." Newton also stated that ."..while

I was examining the debris, Major Marcel was picking up pieces of the target sticks and trying to convince me that some notations on the sticks were alien writings. there were figures on the sticks, lavender or pink in color, appeared to be weather faded markings, with no rhyme or reason (sic). He did not convince me that these were alien writings." Newton concluded his statement by relating that ."..During the ensuing years I have been interviewed by many authors, I have been quoted and misquoted. The facts remain as indicated above. I was not influenced during the original interview, nor today, to provide anything but what I know to be true, that is, the material I saw in General Ramey's office was the remains of a balloon and a RAWIN target."

Balloon Research

The original tasking from GAO noted that the search for information included "weather balloons." Comments about balloons and safety reports have already been made, however the SAF/AAZ research efforts also focused on reviewing historical records involving balloons, since, among other reasons, that was what was officially claimed by the AAF to have been found and recovered in 1947.

As early as February 28, 1994, the AAZD research team found references to balloon tests taking place at Alamogordo AAF (now Holloman AFB) and White Sands during June and July 1947, testing "constant level balloons" and a New York University (NYU)/Watson Labs effort that used "...meteorological devices... suspected for detecting shock waves generated by Soviet nuclear explosions"--a possible indication of a cover story associated with the NYU balloon project. Subsequently, a 1946 HQ AMC memorandum was surfaced, describing the constant altitude balloon project and specified that the scientific data be classified TOP SECRET Priority IA. Its name was Project Mogul (Atch 19).

Project Mogul was a then-sensitive, classified project, whose purpose was to determine the state of Soviet nuclear weapons research. This was the early Cold War period and there was serious concern within the US government about the Soviets developing a weaponized atomic device. Because the Soviet Union's borders were closed, the US Government sought to develop a long range nuclear explosion detection capability. Long range, balloon-borne, low frequency acoustic detection was posed to General Spaatz in 1945 by Dr. Maurice Ewing of Columbia University as a potential solution (atmospheric ducting of low frequency pressure waves had been studied as early as 1900).

As part of the research into this matter, AAZD personnel located and obtained the original study papers and reports of the New York University project. Their- efforts also revealed that some of the individuals involved in Project Mogul were still living. These persons included the NYU constant altitude balloon Director of Research, Dr. Athelstan F. Spilhaus; the Project Engineer, Professor Charles B. Moore; and the military Project Officer, Colonel Albert C. Trakowski .

All of these persons were subsequently interviewed and signed sworn statements about their activities. A copy of theses statements are appended at Atch 20-22. Additionally, transcripts of the interview with Moore and Trakowski are also included (equipment malfunctioned during the interview of Spilhaus) (Atch 23-24). These interviews confirmed that Project Mogul was a compartmented, sensitive effort. The NYU group was responsible for developing constant level balloons and telemetering equipment that would remain at specified altitudes (within the acoustic duct) while a group from Columbia was to develop acoustic sensors. Doctor Spilhaus, Professor Moore, and certain others of the group were aware of the actual purpose of the project, but they did not know of the project nickname at the time. They handled casual inquiries and/or scientific inquiries/papers in terms of "unclassified meteorological or balloon research." Newly hired employees were not made aware that there was anything special or classified about their work; they were told only that their work dealt with meteorological equipment.

An advance ground team, led by Albert P, Crary, preceded the NYU group to Alamogordo AAF, New Mexico, setting up ground sensors and obtaining facilities for the NYU group. Upon their arrival, Professor Moore and his team experimented with various configurations of neoprene balloons; development of balloon "trains" (see illustration, Atch 25); automatic ballast systems- and use of Naval sonobuoys (as the Watson Lab acoustical sensors had not yet arrived). They also launched what they called "service flights." These "service flights" were not logged nor fully accounted for in the published Technical Reports generated as a result of the contract between NYU and Watson Labs. According to Professor Moore, the "service flights" were composed of balloons, radar reflectors and payloads specifically designed to test acoustic sensors (both early sonobuoys and the later Watson Labs devices). The "payload equipment" was expendable and some carried no "REWARD" or "RETURN TO..." tags because there was to be no association between these flights and the logged constant altitude flights which were fully acknowledged. The NYU balloon flights were listed sequentially in their reports (i.e. A,B, 1,5,6,7,8,10...) yet gaps existed for Flights 2-4 and Flight 9. The interview with Professor Moore indicated that these gaps were the unlogged "service flights."

Professor Moore, the on-scene Project Engineer, gave detailed information concerning his team's efforts. He recalled that radar targets were used for tracking balloons because they did not have all the necessary equipment when they first arrived in New Mexico. Some of the early, developmental radar targets were manufactured by a toy or novelty company. These targets were made up of aluminum "foil" or foil-backed paper, balsa wood beams that were coated in an "Elmer's-type" glue to enhance their durability, acetate and/or cloth reinforcing tape, single strand and braided nylon twine, brass eyelets and swivels to form a multi-faced reflector somewhat similar in construction to a box kite (see photographs, Atch 26). Some of these targets were also assembled with purplish-pink tape with symbols on it (see drawing by Moore with Atch 2 1).

According to the log summary (Atch 27) of the NYU group,

Flight A through Flight 7 (November 20, 1946-July 2, 1947) were made with neoprene meteorological balloons (as opposed to the later flights made with polyethylene balloons). Professor Moore stated that the neoprene balloons were susceptible to degradation in the sunlight, turning from a milky white to a dark brown. He described finding remains of balloon trains with reflectors and payloads that had landed in the desert: the ruptured and shredded neoprene would "almost look like dark gray or black flakes or ashes after exposure to the sun for only a few days. The plasticizers and antioxidants in the neoprene would emit a peculiar acrid odor and the balloon material and radar target material would be scattered after returning to earth depending on the surface winds." Upon review of the local newspaper photographs from General Ramey's press conference in 1947 and descriptions in popular books by individuals who supposedly handled the debris recovered on the ranch, Professor Moore opined that the material was most likely the shredded remains of a multi-neoprene balloon train with multiple radar reflectors. The material and a "black box," described by Cavitt, was, in Moore's scientific opinion, most probably from Flight 4, a "service flight" that included a cylindrical metal sonobuoy and portions of a weather instrument housed in a box, which was unlike typical weather radiosondes which were made of cardboard. Additionally, a copy of a professional journal maintained at the time by A.P. Crary, provided to the Air Force by his widow, showed that Flight 4 was launched on June 4, 1947, but was not recovered by the NYU group. It is very probable that this TOP SECRET project balloon train (Flight 4), made up of unclassified components; came to rest some miles northwest of Roswell, NM, became shredded in the surface winds and was ultimately found by the rancher, Brazel, ten days later. This possibility was supported by the observations of Lt Col Cavitt (Atch 17-18), the only living eyewitness to the actual debris field and the material found. Lt Col Cavitt described a small area of debris which appeared, "to resemble bamboo type square sticks one quarter to one half inch square, that were very light, as well as some sort of metallic reflecting material that was also very light ... I remember recognizing this material as being consistent with a weather balloon."

Concerning the initial announcement, "RAAF Captures Flying Disc," research failed to locate any documented evidence as to why that statement was made. However, on July 10, 1947, following the Ramey press conference, the Alamogordo News published an article with photographs demonstrating multiple balloons and targets at the same location as the NYU group operated from at Alamogordo AAF. Professor Moore expressed surprise at seeing this since his, was the only balloon test group in the area. He stated, "It appears that there was some type of umbrella cover story to protect our work with Mogul " Although the Air Force did not find documented evidence that Gen. Ramey was directed to espouse a weather balloon in his press conference, he may have done so because he was either aware of Project Mogul and was trying to deflect interest from it, or he readily perceived the material to be a weather balloon based on the identification from his weather officer, Irving Newton. In either case, the materials recovered by the AAF in July, 1947, were not readily recognizable as anything special (only the purpose was special) and the recovered debris itself was unclassified. Additionally, the press dropped its interest in the matter as quickly as they had jumped on it. Hence, there would be no particular reason to further document what quickly became a "non-event."

The interview with Colonel Trakowski (Atch 23-24) also proved valuable information. Trakowski provided specific details on Project Mogul and described how the security for the program was set up, as he was formerly the TOP SECRET Control Officer for the program. He further related that many of the original radar targets that were produced around the end of World War II were fabricated by toy or novelty companies using a purplish-pink tape with flower and heart symbols on it. Trakowski also recounted a conversation that he had with his friend, and superior military officer in his chain of command, Colonel Marcellus Duffy, in July, 1947. Duffy, formerly had Trakowski's position on Mogul, but had subsequently been transferred to Wright Field. He stated: ."..Colonel Duffy called me on the telephone from Wright Field and gave me a story about a fellow that had come in from New Mexico, woke him up in the middle of the night or some such thing with a handful of debris, and wanted him, Colonel Duffy, to identify it. ... He just said 'it sure looks like some of the stuff you've been launching at Alamogordo and he described it, and I said 'yes, I think it is.' Certainly Colonel Duffy knew enough about radar targets, radiosondes, balloon-borne weather devices. He was intimately familiar with all that apparatus."

Attempts were made to locate Colonel Duffy but it was ascertained that he had died. I-Es widow explained that, although he had amassed a large amount of personal papers relating to his Air Force activities, she had recently disposed of these items. Likewise, it was learned that A.P. Crary was also deceased; however his surviving spouse had a number of his papers from his balloon testing days, including his professional journal from the period in question. She provided the Air Force researchers with this material. It is discussed in more detail within Atch 32. Overall, it helps fill in gaps of the Mogul story.

During the period the Air Force conducted this research, it was discovered that several others had also discovered the possibility that the "Roswell Incident" may have been generated by the recovery of a Project Mogul balloon device. These persons included Professor Charles B. Moore, Robert Todd, and coincidentally, Karl Pflock, a researcher who is married to a staffer who works for Congressman Schiff. Some of these persons provided suggestions as to where documentation might be located in various archives, histories and libraries. A review of Freedom of Information Act (FOIA) requests revealed that Robert Todd, particularly, had become aware of Project Mogul several years ago and had doggedly obtained from the Air Force, through the FOIA, a large amount of material pertaining to it; long before the AAZD researchers independently seized on the same possibility.

Most interestingly, as this report was being written, Pflock published his own report of this matter under the auspices of FUFOR, entitled "Roswell in Perspective" (1994). Pflock concluded from his research that the Brazel Ranch debris originally reported as a "flying disc" was probably debris from a Mogul balloon; however, there was a simultaneous incident that occurred not far away, that caused an alien craft to crash and that the AAF subsequently recovered three alien bodies therefrom. Air Force research did not locate any information to corroborate that this incredible coincidence occurred, however.

In order to provide a more detailed discussion of the specifics of Project Mogul and how it appeared to be directly responsible for the "Roswell Incident," a SAF/AAZD researcher prepared a more detailed discussion on the balloon project which is appended to this report as Atch 32.

Other Research

In the attempt to develop additional information that could help explain this matter, a number of other steps were taken. First, assistance was requested from various museums and other archives (Atch 28) to obtain information and/or examples of the actual balloons and radar targets used in connection with Project Mogul and to correlate them with the various descriptions of wreckage and materials recovered. The blueprints for the "Pilot Balloon Target ML307C/AP Assembly" (generically, the radar target assembly) were located at the Army Signal Corps Museum at Fort Monmouth and obtained. A copy is appended as Atch 29. This blueprint provides the specification for the foil material, tape, wood, eyelets, and string used and the assembly instructions thereto. An actual device was also obtained for study with the assistance of Professor Moore. (The example actually procured was a 1953-manufactured model "C" as compared to the Model B which was in use in 1947. Professor Moore related the differences were minor). An examination of this device revealed it to be simply made of aluminum-colored foil-like material over a stronger paper-like material, attached to balsa wood sticks, affixed with tape, glue, and twine. When opened, the device appears as depicted in Atch 31 (contemporary photo) and Atch 25 (1947 photo, in a "balloon train"). When folded, the device is in a series of triangles, the largest being four feet by two feet ten inches. The smallest triangle section measures two feet by two feet ten inches. (Compare with descriptions provided by Lt Col Cavitt and others, as well as photos of wreckage).

Additionally, the researchers obtained from the Archives of the University of Texas Arlington (UTA), a set of original (i.e. first generation) prints of the photographs taken at the time by the Fort Worth Star-Telegram, that depicted Ramey and Marcel with the wreckage. A close review of these photos (and a set of first generation negatives also subsequently obtained from UTA) revealed several infesting observations. First, although in some of the literature cited above, Marcel allegedly stated that he had his photo taken with the "real" UFO wreckage and then it was subsequently removed and the weather balloon wreckage substituted for it, a comparison shows that the same wreckage appeared in the photos of Marcel and Ramey. The photos also depicted that this material was lying on what appeared

7. AF/SE Memo, March 14, 1994

8. SAF/AQL Memo, March 22, 1994

9. AF/XOWP Memo, March 9, 1994

10. SAF/AAI Memo, March 10, 1994

11. AFHRA/CC Memo, March 8, 1994

12. AFOSI/HO Memo, May 11, 1994

13. List of Locations and Records Searched

14. HQ AAF "Issuance of Orders," June 5, 1947

15. Copy of Vandenberg's Appointment Book and Diary, July 7-9, 1947

16. July 9, 1947 Photos of Balloon Wreckage, Ft Worth Star Telegram

17. Signed Sworn Statement of Cavitt, May 24, 1994

18. Transcript of Cavitt Interview, May 24, 1994

19. Letter, July 8, 1946, Project Mogul

20. Signed Sworn Statement of Spilhaus, June 3, 1994

21. Signed Sworn Statement of Moore, June 8, 1994

22. Signed Sworn Statement of Trakowski, June 29, 1994

23. Transcript of Interview with Moore, June 8, 1994

24. Transcript of Interview with Trakowski, June 29, 1994

25. Illustration of Project Mogul "Balloon Trains"

26. Two Photos of Project Mogul "Balloon Trains"

27. Log Summary, NYU Constant Level Balloon Flights

28. List of Museums Contacted

29. Copy of Blueprint for "Pilot Balloon Target, NM-307C/AP Assembly"

30. Signed Sworn Statement of Newton, July 21, 1994

31. Photos of NM-307C/AP Device, With Vintage Neoprene Balloon and Debris

32. Synopsis of Balloon Research Findings by 1LT James McAndrew

33. "Mensuration Working Paper," With Drawing and Photo

JANAP 146

This page intentionally left blank.

CHAPTER I

GENERAL DESCRIPTION AND PURPOSE OF COMMUNICATION INSTRUCTIONS FOR REPORTING VITAL INTELLIGENCE SIGHTINGS

101. PURPOSE

The purpose of this publication is to provide uniform instructions for reporting of vital intelligence sightings and to provide communication instructions for the passing of these intelligence reports to appropriate military authorities.

102. SCOPE

a. This publication is limited to the reporting of information of vital importance to the security of the United States of America and Canada and their forces, which in the opinion of the observer, requires prompt defensive and/or investigative action by the US and/or Canadian Armed Forces.

b. The procedures contained in this publication are provided for:

(1) US and Canadian civil and commercial aircraft.

(2) US and Canadian government and military aircraft other than those operating under separate reporting directives.

(3) US and Canadian merchant vessels operating under US and Canadian registry.

(4) US and Canadian government and military vessels other than those operating under separate reporting directives.

(5) All other US and Canadian vessels including fishing vessels.

(6) Military installations receiving reports from civilian or military land based or waterborne observers unless operating under separate reporting directives.

102 b. (Continued)

(7) Government and civilian agencies in outlying districts which may initiate reports on receipt of information from land-based observers.

103 MESSAGE IDENTIFICATION

a. Reports made from airborne and landbased sources will be identified by CIRVIS as the first word of the text. (Refer Chapter II)

b. Reports made by waterborne sources will be identified by MERINT as the first word of the text. (Refer Chapters III and IV)

CHAPTER II

CIRVIS REPORTS

SECTION I - GENERAL

201 INFORMATION TO BE REPORTED AND WHEN TO REPORT

a. Sightings within the scope of this chapter, as outlined in Article 102b(1), (2), (6) and (7), are to be reported as follows:

(1) While airborne (except over foreign territory - see paragraph 210) and from land based observers.

NOTE: Canada and the United States are not considered foreign territory for either country for the purposes of this publication.

(a) Single aircraft or formations of aircraft which appear to be directed against the United States or Canada or their forces.

(b) Missiles.

(c) Unidentified flying objects.

(d) Hostile or unidentified submarines.

(e) Hostile or unidentified group or groups of military surface vessels.

(f) Individual surface vessels, submarines, or aircraft of unconventional design, or engaged in suspicious activity or observed in an unusual location or following an unusual course.

(g) Unlisted airfields or facilities, weather stations, or air navigation aids.

(h) Any unexplained or unusual activity which may indicate a possible attack against or through Canada or the United States, including the presence of any unidentified or other suspicious ground parties in the Polar region or other remote or sparsely populated areas.

201\. a. (2) Upon landing

(a) Reports which for any reason could not be transmitted while airborne.

(b) Post-landing reports.

SECTION II - PROCEDURES

202 GENERAL

Communications procedures to be employed will be basically those prescribed for the communications system or service used. Continuing efforts will be made by an aircraft originating a CIRVIS report to insure that each CIRVIS message is received by an appropriate station.

203 PRECEDENCE (PRIORITY OF TRANSMISSION)

Transmission of CIRVIS reports will be preceded by or include the international "Urgency" signal, military precedence of "Emergency", "Rapid US Government", or Canadian "Rush", as appropriate for the communications facilities employed.

EXAMPLE

International Urgency Signal	XXX XXX XXX or PAN PAN PAN
Military Precedence	Y or Emergency
Commercial class of Service Indicator	RAPID US GOVT or RUSH

204 CONTENTS OF CIRVIS REPORTS

a. CIRVIS reports will be similar to a routine aircraft position reports transmitted by either radiotelephone or radiotelegraph. The appropriate procedures to be employed will be those applicable to communications facilities utilized. For the purpose of these examples, military procedures are shown. The reports should contain the following information in the order listed:

(1) CIRVIS REPORT

(2) Identification of reporting aircraft or observer as appropriate.

204 a. (Continued)

(3) Object sighted. Give brief description of the sighting which should contain the following items as appropriate:

(a) Number of aircraft, vessels, missiles, submarines, etc.

(b) Category of object, general description, e.g., size, shape, type of propulsion, etc.

(4) The position of the object. This can be indicated by any of the following methods:

(a) Latitude and longitude.

(b) Over a radio fix.

(c) True bearing and distance from a radio fix.

(d) Over a well-known or well-defined geographic point.

(e) True bearing and distance from a geographic point.

(5) Date and time of sighting (GMT).

(6) Altitude of object.

(7) Direction of travel of object.

(8) Speed of object.

(9) Any observed identification, insignia, or other significant information. Every reasonable effort should be made to positively identify the object sighted.

Example of a radiotelephone transmission:

(Aircraft) PAN PAN PAN - KINDLEY THIS IS AIR FORCE TWO FIVE NINE THREE - CIRVIS REPORT - OVER

(Aeronautical Station) AIR FORCE TWO FIVE NINE THREE THIS IS KINDLEY - GO AHEAD

(Aircraft) EMERGENCY - CIRVIS REPORT - AIR FORCE TWO FIVE NINE THREE SIGHTED FORMATION OF SIX JET BOMBERS - CONFIGURATION IS SWEPT WING WITH EIGHT JET ENGINES - TWO HUNDRED MILES EAST OF BERMUDA ON THIRTEEN MAY AT ONE THREE FIVE ZERO ZEBRA - ALTITUDE THREE FIVE THOUSAND - HEADING TWO SEVEN ZERO DEGREES - NO MARKINGS OBSERVED - OVER

(Aeronautical Station) KINDLEY - ROGER - OUT

Example of a radiotelegraph transmission:

(Aircraft) XXX XXX XXX AFA3 DE A4820

(Aeronautical Station) A4820 DE AFA3 K

(Aircraft) Y - CIRVIS REPORT. A4820 SIGHTED...............ETC.

(Aeronautical Station) A4820 DE AFA3 R AR

205 ADDITIONAL CIRVIS REPORTS

a. Additional reports should be made if more information becomes available concerning a previously sighted object. These reports should contain a reference to the original report sufficient to identify them with the original sighting.

Example of a radiotelephone transmission:

(Aircraft) PAN PAN PAN - KINDLEY THIS IS AIR FORCE TWO FIVE NINE THREE - CIRVIS REPORT - OVER

(Aeronautical Station) AIR FORCE TWO FIVE NINE THREE - THIS IS KINDLEY GO AHEAD

(Aircraft) EMERGENCY - THE SIX JET BOMBERS PREVIOUSLY REPORTED AT ONE THREE FIVE ZERO ZEBRA BY AIR FORCE TWO FIVE NINE THREE ARE NOW ONE THREE ZERO MILES WEST OF BERMUDA AT ONE FOUR THREE FIVE ZEBRA - HEADING TWO SEVEN ZERO DEGREES - OVER

(Aeronautical Station) KINDLEY - ROGER - OUT

Note: In radiotelegraph transmission, the same procedures would apply as prescribed in paragraph 204.

b. Cancellation reports should be made in the event a previously reported sighting is positively identified as friendly. Such reports should be transmitted as a brief message cancelling the previous report(s).

Example of a radiotelephone transmission:

(Aircraft) PAN PAN PAN - KINDLEY THIS IS AIR FORCE TWO FIVE NINE THREE - CIRVIS REPORT - OVER

(Aeronautical Station) AIR FORCE TWO FIVE NINE THREE THIS IS KINDLEY - GO AHEAD

(Aircraft) EMERGENCY - CANCEL CIRVIS REPORT OF ONE THREE FIVE ZERO ZEBRA BY AIR FORCE TWO FIVE NINE THREE - SIX JET BOMBERS POSITIVELY IDENTIFIED AS AIR FORCE BAKER FORTY SEVENS AT ONE FOUR FOUR SIX ZEBRA - OVER

(Aeronautical Station) KINDLEY - ROGER - OUT

Note: In radiotelegraph transmission, the same procedures would apply as prescribed in paragraph 204.

c. A post-landing report is desired immediately after landing to insure receipt of the original report(s) by CINCONAD or RCAF-ADC and to amplify the airborne report(s). This may be filed with either the military or civil communications facility located at the place of landing. If the landing is not made in Canadian or United States territory the report should be made to the nearest Canadian or United States military or diplomatic representative in that area. The post-landing report will refer to the airborne report(s) and, in addition, contain a brief resume of weather conditions at the time of sighting(s), verification of the sighting(s) by other personnel and any other information deemed appropriate.

(1) If no airborne report was made as a result of in-

foreign territory (see paragraph 210), the post-landing report will contain all the information available concerning the sighting.

Example:

EMERGENCY US GOVT

TO: CINCONAD, ENT AFB, COLORADO SPRINGS, COLORADO.

CIRVIS. REFER 281350Z, 281435Z and 281446Z SEPTEMBER 1955 CONCERNING 6 JET BOMBERS TRAVELING DUE WEST OF BERMUDA TOWARD EASTERN COAST OF THE UNITED STATES. AIRCRAFT POSITIVELY IDENTIFIED AS USAF B-47 BY CREW AND PASSENGERS. VISIBILITY UNLIMITED. IDENTIFICATION VERIFIED AT 281457Z. SIGNED MOORE PAA CLIPPER 45V.

206 ADDRESSING

a. Aircraft. It is paramount that all CIRVIS reports reach the appropriate military commands as quickly as possible. The reports, therefore, shall be transmitted as soon as possible after the sighting. Ground procedures have been established to handle CIRVIS reports by either military or civil facilities, so the same procedures as those now established and in use by pilots for air traffic control shall be followed. When contact by civil or military pilots cannot be established with any ground communications station, efforts shall be made to relay the CIRVIS reports via other aircraft with which communication is possible.

(1) Post-landing reports should be addressed to CINCONAD, Ent AFB, Colorado Springs, Colorado, or RCAF Air Defense Command Headquarters, St. Hubert, Quebec, if the sighting occurred within or adjacent to the North American continent. Whichever of these headquarters receives the report will immediately notify the other and also all other addressees of the original report(s). If the sighting(s) occurred in other areas, the post-landing report should be made to the nearest US or Canadian military or diplomatic representative in that area who will forward the report as prescribed in subparagraph 206b(1)(a).

206 (Continued)

b. Communications Stations. Communications stations (to include any civil or military facility such as control tower, naval shore radio station, approach control, ARTC centers, or any other communications facility) receiving CIRVIS reports will immediately after receipting process the report as follows:

(1) US military communications stations will multiple-address the CIRVIS report to the following address designations:

(a) For sightings in overseas areas - reports will be forwarded to:

1. Addresses as prescribed by Area Commanders. (Normally, these addressees are the operating service commands concerned.)

2. Commander-in-Chief, Continental Air Defense Command (CINCONAD), Ent AFB, Colorado Springs, Colorado.

3. Chief of Staff, United States Air Force (C of S, USAF) Washington, D. C.

(b) For sightings within or adjacent to the North American continent, reports will be forwarded to:

1. Commander of the nearest joint air defense division.

2. CINCONAD, Ent AFB, Colorado Springs, Colorado.

3. Appropriate Sea Frontier Command:

a. Commander, Western Sea Frontier (COMWESTSEA-FRON), San Francisco, California.

b. Commander, Eastern Sea Frontier (COMEASTSEA-FRON), New York, N. Y.

4. Chief of Staff, United States Air Force, Washington, D. C.

(2) Canadian military and United States military communications stations in Canada, Alaska and Greenland will multiple-address the CIRVIS report to the following address designations:

(a) For sightings within or adjacent to the North American continent, reports will be forwarded to:

1. RCAF Air Defense Command (CANAIRDEF), St. Hubert, Montreal.

2. Appropriate Flag Officer in Command:

a. Canadian Flag Officer, Atlantic Coast (CANFLAGLANT), Halifax, Nova Scotia.

b. Canadian Flag Officer, Pacific Coast (CONFLAGPAC), Esquimalt, British Columbia.

(3) Civil communications stations will handle CIRVIS reports received from either aircraft or other communications stations as follows:

(a) Air carrier company stations will pass the CIRVIS report, exactly as received, to the nearest CAA or DOT ARTC center in the same manner as air traffic control information.

(b) CAA or DOT communications stations, upon receipt of a CIRVIS report will immediately pass the report to the appropriate ARTC center.

*(c) CAA or DOT ARTC Centers. Upon receipt of CIRVIS reports, ARTC centers will forward them immediately to the appropriate military facility as prescribed by agreement with the appropriate military commander.

*Canadians are to secure agreement regarding this.

207 ACCEPTANCE OF AND RESPONSIBILITY FOR CIRVIS REPORTS.

a. The following activities have responsibilities as follows:

(1) CINCONAD or RCAF-ADC will insure that all CIRVIS reports have been addressed in accordance with

paragraph 206. These headquarters are the normal points of contact between the two countries and are responsible for passing CIRVIS reports of interest, including post-landing reports, to each other.

(2) United States or Canadian military or diplomatic authorities in receipt of CIRVIS reports that have not been previously forwarded will take the action indicated in paragraph 206 without delay by the most rapid means available, giving due consideration to security.

(3) Chief of Staff, USAF, will disseminate CIRVIS reports to appropriate agencies in the Washington, D.C. area.

(4) RCAF-ADC and the Canadian Flag Officers will be responsible for notifying Canadian military headquarters in Ottawa concerning CIRVIS reports.

b. Fixed and mobile military communications facilities and military personnel having occasion to handle CIRVIS reports must lend assistance in all cases required in expediting CIRVIS reports. All civilian facilities and personnel are also urged to do so. Maximum effort must be made by all persons handling CIRVIS reports to insure positive immediate delivery.

c. WHEN A STATION RECEIVES A PARTIAL CIRVIS REPORT AND THE REMAINDER IS NOT IMMEDIATELY FORTHCOMING, IT WILL BE RELAYED OR DELIVERED IN THE SAME MANNER AS A COMPLETE REPORT.

SECTION III - SECURITY

208 MILITARY AND CIVILIAN

Transmission of CIRVIS reports are subject to the Communications Act of 1934, as amended, and the Canadian Radio Act of 1938, as amended. Any person who violates the provisions of these acts may be liable to prosecution thereunder. These reports contain information affecting the National Defense of the United States and Canada. Any person who makes an unauthorized transmission or disclosure of such a report may be liable to prosecution under Title 18 of the US Code, Chapter 37, or the Canadian Official Secrets Act of 1939, as amended.

SECTION IV - EVALUATION REPORTS

209 ACTION BY ACTIVITIES

a. All investigative measures and evaluation processes instituted by addressees and by originating authorities, where applicable, will be handled in accordance with existing procedures and reported in accordance with these instructions, insuring that appropriate commands as listed in paragraph 206 are kept fully informed of investigative results and evaluations. These evaluations shall be expressed in terms indicating the reported sighting as being Positive, Probable, Possible, Improbable or No Threat insofar as being a threat to the security of the United States of America or Canada or their forces, or an explanation of the subject reported when known.

b. The first two words of the text of an evaluation report shall be "CIRVIS EVALUATION" followed by the date-time group and/or other identification of the CIRVIS report(s) being evaluated.

SECTION V - SPECIAL CONSIDERATIONS

210 RADIO TRANSMISSION RESTRICTIONS

CIRVIS reports will not be transmitted by radio while over foreign territory, other than Greenland or Iceland, but will be transmitted as soon as practicable upon leaving foreign territorial boundaries. Foreign territory includes all territory except international water areas and territory under the jurisdiction of the United States of America and Canada.

SECTION VI - COMMERCIAL CHARGES

211 CHARGES.

a. All charges incurred in handling CIRVIS reports through U.S. facilities will be charged to the Department of the Air

Force (accounting symbol "AF"). Insofar as practicable, CIRVIS reports so handled will be forwarded RAPID US GOVT COLLECT.

b. All charges incurred in handling CIRVIS reports through Canadian facilities will be charged to the Royal Canadian Air Force. Insofar as practicable, CIRVIS reports so handled will be forwarded "RUSH COLLECT".

c. Any or all questions of charges will be resolved after traffic has been handled. In no case will CIRVIS reports be delayed because of communication handling charges.

This page intentionally left blank.

Key to the Extraterrestrial Messages

by H. Campaign

This page intentionally left blank.

NSA Technical Journal
Vol. XIV, No. 1

Key To The Extraterrestrial Messages

BY H. CAMPAIGNE

Unclassified

Dr. Campaigne presented a series of 29 messages from outer space in "Extraterrestrial Intelligence," NSA Technical Journal, Vol. XI, No. 2, pp. 101 ff. and in the Special Mathematics and Engineering Issue of the Journal, pp. 117 ff. The following article develops a key to these messages. Paragraph numbers parallel the serial numbers of the messages reprinted in the appendix below. This includes two new series—30 and 31—not included in the previous article.

At every step in the solution we make a guess at the meaning. Evidence will quickly accumulate to verify or refute this guess. The possibility of ambiguity of two consistent solutions is very remote. Only in the last steps, where verification is thin, could this happen.

1. There are 21 symbols, in the order given by this message.

2. B is equivalent to AA, C to AAA, etc. That is, A = 1; B = 2; C = 3; D = 4; E = 5; F = 6; G = 7.

3. The symbol L means the two things that follow are the same. LXY means $x = y$.

4. Each statement has 5 symbols, and begins with L. The 4 symbols after L must be considered as two things. Each statement has a K as the third letter, which must be the start of the second thing. Is B = KAA; C = KBA; C = KAB; D = KCA? If KBA means B + A, it fits.

5. These verify our conclusions on 4. The first means $6 = 1 + (2 + 3)$, the last means $1 + (2 + 3) = (1 + 2) + 3$.

6. Each has five symbols as in 4. They mean 1 = M21; 2 = M31; 1 = M32. Obviously MXY means $x - y$.

7. These translate $N = 1 - 1$; $N = 2 - 2$; $N = 3 - 3$. N stands for zero, 0.

8. These translate 1 = O11; 0 = O01; 2 = O12; 2 = O21; 0 = O02; 0 = O20; 4 = O22, etc. OXY means the product $X \times Y$.

9. These verify the conclusions of 8. The first says that $6 = 1 \times 2 \times 3$, the last $4 \times (5 \times 6) = (4 \times 5) \times 6$.

Note: So far we have seen two kinds of symbols: digits A through G and N, and operators L, K, M, and O. The two digits following the operator are the operands.

10. Translates into 4 = R22; 2 = R21; 1 = R20; 3 = R31; 1 = R30. RXY must mean X^Y, exponentiation. R is another binary operator.

[redacted]

11. Translates into $2^3 = 4+4$; $2^3 = 1+7$; $2^3 = 2\times 4$; $2^2 = 4$, verifying our previous conclusions.

12. Translates into $3^2 = 4+5$; $3^2 = 3\times 3$. Further verification.

Note: In our culture we use parentheses to group closely associated terms, and as a first step it helps, even though it is not necessary, to put in parentheses. To do so unambiguously, start at the right and read left to the first operator symbol; put parentheses about the operator and the two quantities to its right. Repeat until no pair of parentheses contains more than an operator and two quantities.

13. Translates into $J = 2^3$; $J = 1+7$; $J+1 = 3\times 3$; $J+1 = 3^2$, therefore $J = 8$.

14. We can only introduce parentheses by assuming P is an operator, so we get 2 = P84; 4 = P82; 1 = P33; and 3 = P62. Thus $P \times Y = X \div Y$, division.

15. Assume U is an operator, getting U12; U23; U34; U58; U68; U78; $U2^3 3^2 = U89$. The smaller is first in each case; so perhaps UXY means X precedes Y or $X < Y$.

16. The new character Q must be an operator. Transcribed it gives us Q: $0=1$; Q: $1=3$; Q: $1=1+1$; Q: $0=1\times 1$; Q: $2^2=2$; Q: $2<2$; Q: H$(1=1)$ $(1=2)$; Q|Q: $3=3$|; Q: $8<7$.

Clearly Q means "the following statement is false." Then the next to the last is read "it is false that $3 \neq 3$." QL will be translated $\neq$. The second new symbol is not clear, except that it is an operator whose operands are statements, not quantities, a Boolean operator.

Note: Q is an operator with only one operand, unary.

17. Putting in parentheses shows that S is also a unary operator operating on statements. Transcribed they are: S: $1=1$; S: $1<2$; S: $2=1+1$; S: $1=1\times 1$; S: $0=1-1$; S: H|$(6=1\times 6)$ $(1=6\div 6)$|; S: H|$(1=1)$ $(2=2)$|. It is apparent that S means "the following statement is true" or "it is asserted that." The next to the last message shows that HXY means "X implies Y" or "X is a consequence of Y" or maybe "X is logically equivalent to Y."

18. Our rule for parentheses breaks down unless T is a different kind of symbol. The first message shows that T may be a unary operator on quantities, so that AT or TA is a quantity. The third message shows that it must be the first, since T is last. Putting in parentheses this way gives 1T = 1; 2T = 1T + 1; 5T = 2T + 3T; 6T = 3T × 2T; 7T = 7; 10T = 8. T must be an ending. On one digit it makes no difference. It combines the two digits 10 to make 8. Octal arithmetic?

19. Transcribes to $123T = 1\times 8^2 + (2\times 8^1 + 3)$;
$321T = 3\times 8^2 + (2\times 8^1 + 1\times 8^0)$;
$4567T = (4\times 8^3 + 5\times 8^2) + (6\times 8^1 + 7\times 8^0)$.

Clearly T indicates that "the preceding digits form an octal number." Possibly it is an octal point; if so, digits may occur after it.

Note: Because of the way grouping is implied, it is sufficient to have a marker at the end of a number in order to clearly isolate it as a single entity.

20. In trying to put on parentheses it appears that V is also an ending. But this one combines with both quantities (that is, digits.) and operator. Transcribing by treating V and the preceding symbol as a single unit for the time being, we get:

8 $\quad$ 1 + AV implies 7 = AV.

I will use [illegible] · for H hereafter. Remember that we are not sure of the sense of this sign. 11 [illegible] - 3'' = · 2 = AV (I have omitted the T. Remember that 11T is nine);

3 $\quad$ AV = · 11 = $\overline{AV}^2$;

5 $\quad$ AV [illegible] · 6 · 1 + AV.

In the next message if we combine the O and V into one symbol the message does not parse. Try GOV as one symbol, getting

3 $\quad$ 1 + GOV = · 2 = GOV;

6 $\quad$ 2 × SOV = · 3 = SOV;

3 $\quad$ DOV = · 11 = $\overline{DOV}^2$;

It is true that AV + PV = PV + AV;

It is a tautology that AV × PV = PV × AV;

It is an identity that AV + (PV + TV) = (AV + PV) + TV;

It is asserted that AV × (PV × TV) = (AV + PV) + TV;

AV ≠ BV = · AV - BV ≠ BV [illegible] AV;

AV ≠ BV [illegible] · AV'' ≠ BV''.

The meaning of V must be that "the preceding letters as a group have an abstract meaning, or are a variable." V is a little like a word spacer.

Note: Putting in parentheses is now complicated by another rule. Each T or V should be packaged with preceding symbols, just how many depending on the parsing of the message. Those preceding T will all be digits. Those preceding V can be expected to reoccur as a group.

21. Putting parentheses in these messages is difficult until we notice that UV appears in each. They then transcribe into:

0 $\quad$ |2 + ($\overline{D}\overline{D}\overline{D}V^2$ − DDDV × 3)| = · UV|1 = DDDV| |2 = DDDV|;

UV|1 = DV||0 − 1 = DV| = · |1 = $\overline{DV}^2$|;

UV|1 = BV||0 = BV| = · 0 = BV − $\overline{BV}^2$;

It is true that UV|AV = BV||AV ≠ BV|;

It is true that UV|AV < BV||BV < AV|.

In order to complete the parsing we had to assume that UV was a binary operator, and in every case the operands are statements. It is clear from the algebra that UV means "or." The last message shows that U means ≤, rather than < as I had it.

22. We notice that TV is used in every message, and parallels the usage of UV. Assuming TV is a binary Boolean operator, the messages parse.

It is false that TV|AV ≤ BV||BV ≤ AV|;
It is false that TV|AV = BV||AV ≠ BV|;
TV|$\bar{\text{AV}}^2$ = 4||0 ≤ AV| = · AV = 2;
AV > BV = · |BV ≤ AV| or |BV = AV|;
It is true that not TV GV HV = · GV or HV;
It is true that GV and HV = · TV GV HV;
AV or (BV or CV) = (AV or BV) or CV;
TV|AV TV|BVCV|| = TV||TV AV BV|CV|;
TV|AV|BV or CV|| = TV|AVBV| or TV|AV CV|;
AV or TV |BV CV| = TV|AV or BV||AV or CV|.

It is apparent that TV means "and." Notice that L is used here to mean "logically equivalent to," although I have written "=".

Note: U is used here for <, not ≤.

Either there is a mistake, or the usage varies.

23. The parsing falters until we realize that JNV occurs in each message, and is probably a word. BAV and CAV also occur in each message. They transcribe into:

JNV |BAV or CAV| BAV;
JNV BAV |BAV and CAV|;
JNV |BAV or CAV||BAV and CAV|;
JNV BAV CAV = · BAV = (BAV or CAV);
JNV BAV CAV = · CAV = (BAV and CAV).

The last two conclusions look like set theory statements. JNV parses like a binary operator. JNV XY could mean "X contains Y" in the set theory sense. Then if UV is "or" in the set theory sense, the union, and TV is "and" in the set theory sense, the intersection, the statements above can be rewritten:

BAV ∪ CAV ⊃ BAV
BAV ⊃ BAV ∩ CAV
BAV ∪ CAV ⊃ BAV ∩ CAV
BAV ⊃ CAV = · BAV = (BAV ∪ CAV)
BAV ⊃ CAV = · CAV = (BAV ∩ CAV).

24. NKV looks like a binary operator of which at least the first operand is a quantity. JAV is uniformly the second operand. From 23 above we are alert to set theory statements. Could it be that NKV says something is a member of some set? Try it. They become

1∈JAV; 2∈JAV; 3∈JAV; 4∈JAV; 5∈JAV; 6∈JAV; 7∈JAV; 11∈JAV; 12∈JAV; AV∈JAV = · AV + 1∈JAV.

JAV is the set of positive integers! It fits!

25. These parse into:

(1 and 2) ε JAV;
((1 and 2) and 3) ε JAV;
(14 and 17) ε JAV;
((77 and 100) and 101) ε JAV;
(AV ⊃ NMV) and (BV ⊃ NMV) =· (AV and BV) ⊃ NMV;
 0 ε JAV;
8 ε JAV; $8'$ ε JAV; $8'$ ε JAV; $8'$ ε JAV; $8''$ ε JAV; $8'''$ ε JAV; $8''''$ ε JAV;
$8'''''$ ε JAV; BV ε JAV =· 8^{BV} ε JAV;
(BV and CV) ε JAV =· BV + CV ε JAV;
(BV and CV) ε JAV =· BV × CV ε JAV;
(BV and CV) ε JAV =· BV^{CV} ε JAV;
1/2 ε JAV; 1 − 2 ε JAV; 0 − 3 ε JAV; 7/6 ε JAV.

This verifies beyond doubt the guess of 24.

26. There is a new word, JOV. The messages read JAV ε JOV:

0 − 1 ε JOV; 0 − BV ε JAV =· BV ε JOV;
1/2 ε JOV; AV and BV ε JOV =· AV − BV ε JOV;
(AV and BV in JOV) and 0 ≠ BV =· AV ÷ BV in JOV;
1 ÷ 0 not in JAV; 1 ÷ 0 not in JOV;
It is true that (AV ÷ BV) × (CV ÷ DV) = (AV × CV) ÷ (BV × DV);
It is true that AV × DV < BV × CV =· AV ÷ BV < CV ÷ DV, BV × DV ≠ 0;
AV ε JAV =· 0 − 1 < AV.

JOV is seen to be the field generated by JAV, in other words, the set of rational numbers. The next to the last message has a garble, an extraneous A.

27. This transcribes to:

(AV =· BV) and (BV =· AV) =· HV.AV.BV.
Clearly HV means "logically equivalent," or "·=·".
(AV ·=· BV) =· (AV =· BV) and (BV =· AV)
(AV ·=· BV) ·=· (AV =· BV) and (BV =· AV).

28. These transcribe to

$\overline{GV^2} = 3$ =· GV not in JOV;
$\overline{GV^2} = 2$ =· GV not in JOV;
$\overline{GV^2} = 5$ =· GV not in JOV;
$GV^2 = 5$ =· GV in JEV;
 JOV is in JEV;
 JAV is in JEV;
$\overline{GV^2} = 0 - 1$ =· GV not in JEV.

We have a new set, containing the rationals, and at least one irrational, but not the imaginary $\sqrt{-1}$. JEV is probably the real numbers.

29. These transcribe to

$1 - 2^{-1}$ in JBV; $1 - 3^{-1}$ in JBV;

$1 - 4^{-1}$ in JBV; $1 - \overline{NV}^{-1}$ in JBV;

NLV JBV = 1 [assuming NLV is one word. Another possible parsing is LV(JBV, 1) = 0]

1/1 in JCV; 1/2 in JCV; 1/3 in JCV; 1/NV in JCV; NLV JCV = 0 [or LV (JCV, 0) = 0. But the two examples suggest that NLV means "a limit of." If NV is an integer this fits perfectly.]

$(1 - 1/2)^2$ in JBV; $(1 - 1/3)^3$ in JBV;

$(1 - 1/4)^4$ in JBV; $\left(1 - \frac{1}{8^{100}}\right)^{8^{100}}$ in JBV;

$\left(1 - \frac{1}{NV}\right)^{NV}$ in JBV; NLV JBV in JEV.

If NLV means limit, then JEV contains the number e, a verification of our guess that JEV named the real numbers.

The last two lessons—30 and 31—were not published with the first twenty-nine because it made too long an exercise.

30. The later messages of this group have the mysterious sequences ABCD, ABCDE, DEFG, etc, each ending with STV. If we bunch these each as a unit, the messages parse. They then say JNV 1 natural number; JNV 2 natural number; JNV 3 natural number; JNV 123 STV natural number; conjecture STV means "the preceding is a set (or sequence)," and JNV means "belongs to." There is doubt about the latter, since we thought earlier that it meant "contains"; AV belongs to 1234 = · AV is a natural number; 12345 or 4567 = 45 as sets: 12345 and 4567 = 1234567 as sets.

31. This last group is of impressive magnitude, 41 messages, of which the thirtieth is quite long. Parsing is eased by the parallel construction of the messages. They transcribe to:

JRAV belongs to CHAV; JRBV belongs to CHAV; JRGV belongs to CHAV; the set JRAV, JRBV, JRCV, JRDV, JREV, JRFV, JRGV belongs to CHAV; [Since all the digits appear in these groups, maybe they are used like subscripts and should be read JR_1, JR_2, etc.]; JO_1 belongs to CHAV; JO_2 belongs to CHAV; JO_{22} belongs to CHAV; the set JO_1, JO_2, JO_3, JO_4, JO_5, JO_6, JO_7, JO_{10}, JO_{11}, JO_{12}, JO_{13}, JO_{14}, JO_{15}, JO_{16}, JO_{17}, JO_{20}, JO_{21}, JO_{22} belongs to CHAV; U_{10} and U_1 = · 22 JO_1 belongs to CHAV [This one must be parsed wrong or garbled]; BL_1 belongs to JR_1; BL_2 belongs to JR_1; BL_3 belongs to JR_2; BL_4 belongs to JR_2; BL_{12} belongs to JR_2;

AV ≥ 3 and 12 ≥ AV · = · BL_{AV} belongs to JR_3;

AV ≥ 13 and 22 ≥ AV · = · BL_{AV} belongs to JR_4;

AV ≥ 23 and 44 ≥ AV · = · BL_{AV} belongs to JR_5;

AV > 45 and 66 > AV ⇒ BL[illegible] belongs to JR_5;
AV > 67 and 126 > AV ⇒ BL[illegible] belongs to JR_6;
AV > 127 and 142 > AV ⇒ BL[illegible] belongs to JR_7;
The set BL[illegible], BL[illegible], BL[illegible], BL[illegible], BL[illegible], BL[illegible], BL[illegible] belongs to JO_1;
The set BL[illegible], BL[illegible], BL[illegible], BL[illegible], BL[illegible], BL[illegible], BL[illegible] belongs to JO_2;
The set BL[illegible], BL[illegible], BL[illegible], BL[illegible], BL[illegible] belongs to JO_3;
The set BL[illegible] and BL[illegible] and BL[illegible] and BL[illegible] and BL[illegible] belongs to JO_4;
The set BL[illegible], BL[illegible], BL[illegible], BL[illegible], BL[illegible] belongs to JO_5;
The set BL[illegible], BL[illegible], BL[illegible], BL[illegible], BL[illegible] belongs to JO_6 [note a garble here, an N is repeated];
The set BL[illegible], BL[illegible], BL[illegible], BL[illegible], BL[illegible] belongs to JO_7;
The set BL[illegible], BL[illegible], BL[illegible], BL[illegible], BL[illegible] belongs to JO_{10};
The set BL[illegible], BL[illegible] belongs to JO_{11};
BL[illegible] and BL[illegible] and BL[illegible] belongs to JO_{12};
The set BL[illegible], BL[illegible], BL[illegible] belongs to JO_{13};
The set BL[illegible], BL[illegible], BL[illegible] belongs to JO_{14};
The set BL[illegible], BL[illegible], BL[illegible] belongs to JO_{15};
The set BL[illegible], BL[illegible], BL[illegible] belongs to JO_{16};
The set BL[illegible], BL[illegible], BL[illegible] belongs to JO_{17};
The set BL[illegible], BL[illegible], BL[illegible] belongs to JO_{20};
The set BL[illegible], BL[illegible], BL[illegible] belongs to JO_{21};
The set BL[illegible], BL[illegible], BL[illegible] belongs to JO_{22};
CHAV belongs to KSPV.

The transcription leaves a lot to be resolved. There are several words the meanings of which are yet to be determined. The word CHAV (or CH,) seems to be central. There are seven words JR, and eighteen words JO, and each of these belongs to CHAV. There are 98 words BL, each of which seems to belong to a unique JO,. Does each also belong to a unique JR,? With this hint we can straighten out the garbled message above; it reads "0 < AV and AV < 22 ⇒ JO_{11} belongs to CHAV"; there was a V omitted. I was also able to reparse six other messages. I will not bore you with the details, since the list above has been corrected.

Since each BL belongs to one JR, and JO,, these can be displayed in a matrix

	JR_1	JR_2	JR_3	JR_4	JR_5	JR_6	JR_7
JO_1	BL[illegible]	BL[illegible]	BL[illegible]	BL[illegible]	BL[illegible]	BL[illegible]	BL_{127}
JO_2	BL[illegible]	BL[illegible]	BL[illegible]	BL[illegible]	BL[illegible]	BL_{70}	BL[illegible]
JO_3		BL[illegible]	BL[illegible]	BL[illegible]	BL[illegible]	BL[illegible]	
JO_4		BL[illegible]	BL[illegible]	BL[illegible]	BL[illegible]	BL_{122}	
JO_5		BL[illegible]	BL[illegible]	BL[illegible]	BL[illegible]	BL_{123}	
JO_6		BL[illegible]	BL[illegible]	BL_{42}	BL[illegible]	BL_{124}	

JO_7	BL_{11}	BL_{21}	BL_{43}	BL_{65}	BL_{125}					
JO_{10}	BL_{12}	BL_{22}	BL_{41}	BL_{66}	BL_{126}					
JO_{11}	BL	25	47	71	72	73	74	75	76	77
	100	101	102	103	104	105	106	107		
		131	132	133	134	135	136	137		
	140	141	142							
JO_{12}	BL_{16}			BL_{50}	BL_{110}					
JO_{13}			BL_{27}	BL_{51}	BL_{111}					
JO_{14}			BL_{30}	BL_{52}	BL_{112}					
JO_{15}			BL_{31}	BL_{53}	BL_{113}					
JO_{16}			BL_{32}	BL_{54}	BL_{114}					
JO_{17}			BL_{33}	BL_{55}	BL_{115}					
JO_{20}			BL_{34}	BL_{56}	BL_{116}					
JO_{21}			BL_{35}	BL_{57}	BL_{117}					
JO_{22}			BL_{36}	BL_{60}	BL_{120}					

Remember that these are not decimal numbers. There is only one cell with more than one entry, and the subscripts in it in decimal notation are 21, 39, 57, 58, 59, 60, 61, 62, 63, 64, 65, 66, 67, 68, 69, 70, 71, 89, 90, 91, 92, 93, 94, 95, 96, 97, 98. The larger part of the entries is systematically distributed in the first eight rows. This suggests the periodic table of the chemical elements! On consulting a table we find, sure enough, that elements 57 through 71 are rare earths, and are lumped into one cell. Some, but not all, authorities also list 89 through 103 as rare earths. Elements 21 and 39 are Scandium and Yttrium.

CHAV must mean the periodic table. JR_x means column x, and JO_y means row y. BL_z means element Z. The meaning of KSPV is not known, except that it is a generalization of "periodic table." It may merely mean table, or scientific fact, or university subject.

Looking back over the exercise we see we have penetrated the meaning of the basic symbols, and even more important, have learned some of the syntax rules of the notation, and have caught mistakes in the process. We have a few words for sophisticated concepts, and, given more data, with a little labor we could establish its translation.

The concepts used here are the basic ones of number, sets, and physical constants which any cultures must share. How bizarre the syntax and values of a culture could be I cannot conjecture, but any civilizations capable of sending a message across space must have many things in common.

APPENDIX

Recently a series of radio messages was heard coming from outer space. The transmission was not continuous, but cut by pauses into pieces which could be taken as units, for they were repeated over and over again. The pauses show here as punctuation. The various combinations have been represented by letters of the alphabet, so that the messages can be written down. Each message except the first is given here only once. The serial number of the message has been supplied for each reference.

1 ABCDEFGHIJKLMNOPQRSTUV
ABCDEFGHIJKLMNOPQRSTUV
ABCDEFGHIJKLMNOPQR etc

2 AA. B AAA.C. AAAA.D. AAAAA.E. AAAAAA.F. AAAAAAA.G

3 LAA. LBB. LCC. LDD. LEE. LFF. LGG

4 LBKAA. LCKBA. LCKAB. LDKCA. LDKBB. LDKAC: LEKDA.
LEKCB. LEKBC. LEKAD. LFKEA. LFKDB. LFKCC. LFKBD.
LFKAE - -

5 LFKAKBC LFKCKBA. LGKAKBD. LGKCKAC. LKAKBCKKABC

6 LAMBA. LBMCA. LAMCB. LCMDA. LBMDB. LAMDC. LDMEA.
LCMEB. LBMEC. LAMED

7 LNMAA. LNMBB. LNMCC. LNMDD. LNMEE. LNMFF. LNMGG

8 LAOAA. LNONA. LBOAB. LBOBA. LNONB. LNOBN. LDOBB:
LDOAD. LFOAF. LFOBC

9 LFOAOBC. LFOCOBA. LFOBOCA; LODOEFOODEF

10 LDRBB LBRBA. LARBN. LCRCA. LARCN

11 LRBCKDD. LRBCKAG. LRBCOBD. LRBBD

12 LRCBKDE. LRCBOCC

13 LJRBC. LJKAG. LKJAOCC. LKJARCB

14 LBPJD. LDPJB. LAPCC. LCPFB

15 UAB. UBC; UCD; UEJ; UFJ; UGJ; URBCRCB.

16 QLNA; QLAC; QLAKAA; QLNOAA; QLRBBB; QUBB; QHLAALAB;
QQLCC; QUJG.

17 SLAA; SUAB; SLBKAA; SLAOAA; SLNMAA; SHLFOAFLAPFF; SHLAALBB

18 LATA. LBTKATA; LETKBTCT; LFTOCTBT; LGTG; LANTJ.

19 LABCTKOARJBKOBRJAC;
LCBATKOCRJBKOBRJAOARJN;
LDEFGTKKODRJCOERJBKOFRJAOGRJN;

20 HLJKAAVLGAV.
HLAATHCAVLBAV.
HLCAVLAPTHAVB
HLEAVLFKAAV
HLCKAGJVLBCOV.
HLFGBSOVLCSOV
HLCTQVLAATHDOVB
SLKAVFSKFVAV
SLOAVTVOPVAV.
SLKAVKPVTVKKAVPVTV.
SLOAVOPVSVOOAVPVSV.
HQLAVBVQLMAVBVMBVAV.
HQLAVBVQLRAVBVRBVAV

21 HLNKNMRDODVBODODVCUVLADODVLBODDV
HUVLADVLMNADVLAROVB
HUVLABVLNBVLMMBVRBVB.

SUVLAVBVQLAVBV.
SIIVUAVBVUBVAV

22 QTVUAVBVUBVAV.
QTVLAVBVQLAVBV.
HTVLRAVBNUNAVLAVB.
HQUAVBVUVUBVAVLBVAV.
SHQTVGVHVUVQGVQHV.
SHQUVGVHVTVQGVQHV.
LUVAVUVBVCVHVUVAVBVCV.
LTVAVTVBVCVTVTVAVBVCV.
LTVAVUVBVCVHVTVAVBVTVAVCV.
LUVAVTVBVCVTVUVAVBVUVAVCV

23 JNVUVHAVCAVBAV.
JNVBAVTVBAVCAV.
JNVUVBAVCAVTVBAVCAV.
HJNVBAVCAVLBAVUVBAVCAV.
HJNVBAVCAVLCAVTVBAVCAV.
HTVJNVBAVCAVJNVCAVBAVLBAVCAV.
HJNVBAVCAVJNVQCAVQBAV.
SHTVJNVBAVCAVJNVCAVDAVJNVBAVDAV

24 NKVAJAV.
NKVBJAV.
NKVCJAV.
NKVDJAV.
NKVEJAV.
NKVFJAV.
NKVGJAV.
NKVAATJAV.
NKVABTJAV.
HNKVAVJAVNKVKAVAJAV

25 NKVTVABJAV.
NKVTVTVABCJAV.
NKVTVADTAGTJAV.
NKVTVTVGGTANNTANATJAV.
HTVJNVAVNMVJNVBVNMVJNVTVAVBVNMV.
NKVHJAV.
NKVJJAV.
NKVRJBJAV.
NKVRJCJAV.
NKVRJIJJAV.
NKVRJJJAV.
NKVRJANTJAV.
NKVRJANNTJAV
NKVRJANNNTJAV
HNKVRVJAVNKVRJBVJAV

HNKVTVBVCVJAVHNVHBVCVJAV
HNKVTVBVCVJAVNKVUBVCVJAV
HNKVTVBVCVJAVNKVRBVCVJAV.
QNKVPABJAV.
QNKVHABJAV.
QNKVMNCJAV.
QNKVPGFJAV

26 NKVJAVJOV.
NKVMNAJOV.
HNKVMNBVJAVNKVBVJOV.
NKVPABJOV.
HNKVTVAVBVJOVNKVMAVBVJOV.
HTVNKVTVAVBVJOVQLNBVNKVPAVBVJOV.
QNKVPANJAV.
QNKVPANJOV.
SLOPAVBVPCVUVPUAVCVOBVDV
SHHOAVDVOHVCVHPAVBVPACVDV. QLOBVDVN
HNKVAVJAVUMNAAV.

27 HTVHAVBVHBVAVHVAVBV.
HHVAVBVTVHAVBVHBVAV
HVHVAVBVTVHAVBVHBVAV

28 HLRGVBCQNKVGVJOV.
HLRGVBBQNKVGVJOV
HLRGVBEQNKVGVJOV.
HLRGVBENKVGVJEV.
NKVJOVJEV.
NKVJAVJEV.
HLRGVBMNAQNKVGVJEV

29 NKVMARBMNAJBV.
NKVMARCMNAJBV.
NKVMARDMNAJBV.
NKVMARNVMNAJBV
LNLVJBVA.
NKVPAAJCV.
NKVPABJCV.
NKVPACJCV.
NKVPANVJCV.
LNLVJCVN
NKVRMAPABBJBV.
NKVRMAPACCJBV.
NKVRMAPADDJBV.
NKVRMAPARJANNTRJANNTJBV.
NKVRMAPANVNVJBV.
NKVRLVJBVJEV.

30. JNVAJAV;
JNVBJAV;
JNVCJAV;
JNVABCSTVJAV;
HJNVAVABCDSTVJNVAVJAV;
LUVABCDESTVDEFGSTVDESTV;
LTVABCDESTVDEFGSTVABCDEFGSTV;

31. JNVJRAVCHAV;
JNVJRBVCHAV;
JNVJRGVCHAV;
JNVJRAVJRBVJRCVJRDVJREVJRFVJRGVSTVCHAV.
JNVJOAVCHAV;
JNVJOBVCHAV;
JNVJOBBVCHAV;

```
.JNV.JOAV.JOBV.JOCV.JODV.JOEV.JOFV.JOGV
.JOANV.JOAAV.JOABV.JOACV.JOADV.JOAEV.JOAFV.JOAGV
.JOBNV.JOBAV.JOBBVSTVCHAV ;
HTVUNAVUAVBBT.JNV.JOAVVCHAV ;
.JNVBL.AV.JRAV .
.JNVBL.BV.JRAV .
.JNVBL.CV.JRBV
.JNVBL.DV.JRBV
.JNVBL.ABV.JRBV .
HVTVQUAVCQUABTAV.JNVBL.AVV.JRBV .
HVTVQUAVACTQUBBTAV.JNVBL.AVV.JRCV
HVTVQUAVBCTQUDDTAV.JNVBL.AVV.JRDV .
HVTVQAVDETQUFFTAV.JNVBL.AVV.JREV .
HVTVQUAVFGTQUABFTAV.JNVBL.AVV.JRFV .
HVTVQUAVABCTQUADBTAV.JNVBL.AVV.JRGV ;
.JNVBL.AVBL.CVBL.ACVBL.BCVBL.DEVBL.FGV
BL.ABGVSTV.JOAV .
.JNVBL.BVBL.DVBL.ADVBL.BDVBL.DFV
BL.GNVBL.ACNVSTV.JOBV ;
.JNVBL.EVBL.AEVBL.CGVBL.FAVBL.ABAVSTV.JOCV .
.JNVTVTVTVTVBL.FVBL.AFVBL.DNVBL.FBVBL.ABBVSTV.JODV .
.JNVBL.GVBL.ACVBL.DAVBL.FCVBL.ABCVSTV.JOEV .
.JNVBL.ANNBL.BNVBL.DBVBL.FDVBL.ABDVSTV.JOFV .
.JNVBL.AAVBL.BAVBL.DCVBL.FEVBL.ABEVSTV.JOGV ;
.JNVBL.ABVBL.BBVBL.DDVBL.FFVBL.ABFVSTV.JOANV .
.JNVBL.BEVBL.DGVBL.GAVBL.GBVBL.GCVBL.GDV
BL.GEVBL.GFVBL.GGVBL.ANNVBL.ANAVBL.ANBVBL.ANCV
BL.ANDVBL.ANEVBL.ANFVBL.ANGVBL.ACAVBL.ACBV
BL.ACCVBL.ACDVBL.ACEVBL.ACFVBL.ACGVBL.ADNV
BL.ADAVBL.ADBVSTV.JOAAV ;
.JNVTVTVBL.BFVBL.ENVBL.AANV.JOABV .
.JNVBL.BGVBL.EAVBL.AAAVSTV.JOACV ;
.JNVBL.CNVBL.EBVBL.AABVSTV.JOADV .
.JNVBL.CAVBL.ECVBL.AACVSTV.JOAEV ;
.JNVBL.CBVBL.EDVBL.AADVSTV.JOAFV .
.JNVBL.CCVBL.EEVBL.AAEVSTV.JOAGV ;
.JNVBL.CDVBL.EFVBL.AAFVSTV.JOBNV ;
.JNVBL.CEVBL.EGVBL.AAGVSTV.JOBAV ;
.JNVBL.CFVBL.FNVBL.ABNVSTV.JOBBV ;
.JNVCHAVKSPV
```

This page intentionally left blank.

Extraterrestrial Intelligence

by Howard H. Campaigne

This page intentionally left blank.

Extraterrestrial Intelligence

BY HOWARD H. CAMPAIGNE

Unclassified

Extraterrestrial Communications

In the most recent issue of the NSA Technical Journal–Vol. XI, No. 1–Mr. Lambros D. Callimahos discussed certain aspects of extraterrestrial intelligence and included several messages to test the reader's ingenuity. In the following pages, Dr. H. H. Campaigne offers additional communications from outer space.

Recently a series of radio messages was heard coming from outer space. The transmission was not continuous but was cut by pauses into pieces which could be taken as units, for they were repeated over and over again. The pauses show here as punctuation. The various combinations have been represented by letters of the alphabet, so that the messages can be written down. Each message except the first is given here only once. The serial number of the message has been supplied for each reference.

1. ABCDEFGHJKLMNOPQRSTUV
   ABCDEFGHJKLMNOPQRSTUV
   ABCDEFGHJKLMNOPQR etc.
2. AA, B;AAA,C; AAAA,D; AAAAA,E; AAAAAA,F; AAAAAAA,G.
3. LAA; LBB, LCC; LDD; LEE; LFF; LGG.
4. LBKAA; LCKBA; LCKAB; LDKCA; LDKBB; LDKAC; LEKDA; LEKCB; LEKBC; LEKAD; LFKEA; LFKDB; LFKCC; LFKBD; LFKAE.
5. LFKAKBC; LFKCKBA; LGKAKBD; LGKCKAC; LKAKBCKKABC.
6. LAMBA; LBMCA; LAMCB; LCMDA; LBMDB; LAMDC; LDMEA; LCMEB; LBMEC; LAMED.
7. LNMAA; LNMBB; LNMCC; LNMDD; LNMEE; LNMFF; LNMGG.
8. LAOAA; LNONA; LBOAB; LBOBA; LNONB; LNOBN; LDOBB; LDOAD; LFOAF; LFOBC.
9. LFOAOBC; LFOCOBA; LFOBOCA; LODOEFOODEF.
10. LDRBB; LBRBA; LARBN; LCRCA; LARCN.
11. LRBCKDD; LRBCKAG; LRBCOBD; LRBBD.
12. LRCBKDE; LRCBOCC.
13. LJRBC; LJKAG; LKJAOCC; LKJARCB.
14. LBPJD; LDPJB; LAPCC; LCPFB.

15. UAB; UBC; UCD; UEJ; UFJ; UGJ; URBCRCB.

16. QLNA; QLAC; QLAKAA; QLNOAA; QLRBBB; QUBB; QHLAALAB; QQLCC; QUJG.

17. SLAA; SUAB; SLBKAA; SLAOAA; SLNMAA; SHLFOAFLAPFF; SHLAALBB.

18. LATA; LBTKATA; LETKBTCT; LFTOCTBT; LGTG; LANTJ.

19. LABCTKOARJBKOBRJAC;
LCBATKOCRJBKOBRJAOARJN;
LDEFGTKKODRJCOERJBKOFRJAOGRJN;

20. HLJKAAVLGAV;
HLAATRCAVLBAV;
HLCAVLAATRAVB;
HLEAVLFKAAV;
HLCKAGOVLBGOV;
HLFOBSOVLCSOV;
HLCDOVLAATRDOVB;
SLKAVPVKPVAV;
SLOAVPVOPVAV;
SLKAVKPVTVKKAVPVTV;
SLOAVOPVSVOOAVPVSV;
HQLAVBVQLMAVBVMBVAV;
HQLAVBVQLRAVBVRBVAV;

21. HLNKBMRDDDVBODDDVCUVLADDDVLBDDDV;
HUVLADVLMNADVLARDVB;
HUVLABVLNBVLNMBVRBVB;
SUVLAVBVQLAVBV;
SUVUAVBVUBVAV.

22. QTVUAVBVUBVAV;
QTVLAVBVQLAVBV;
HTVLRAVBDUNAVLAVB;
HQUAVBVUVUBVAVLBVAV;
SHQTVGVHVUVQGVQHV;
SHQUVGVHVTVQGVQHV;
LUVAVUVBVCVUVUVAVBVCV;
LTVAVTVBVCVTVTVAVBVCV;
LTVAVUVBVCVUVTVAVBVTVAVCV;
LUVAVTVBVCVTVUVAVBVUVAVCV.

23. JNVUVBAVCAVBAV;
JNVBAVTVBAVCAV;
JNVUVBAVCAVTVBAVCAV;
HJNVBAVCAVLBAVUVBAVCAV;
HJNVBAVCAVLCAVTVBAVCAV;
HTVJNVBAVCAVJNVCAVBAVLBAVCAV;
HJNVBAVCAVJNVQCAVQBAV;
SHTVJNVBAVCAVJNVCAVDAVJNVBAVDAV.

24. NKVAJAV;
NKVBJAV;
NKVCJAV;
NKVDJAV;
NKVEJAV;
NKVFJAV;
NKVGJAV;
NKVAATJAV;
NKVABTJAV;
HNKVAVJAVNKVKAVAJAV.

25. NKVTVABJAV;
NKVTVTVABCJAV;
NKVTVADTAGTJAV;
NKVTVTVGGTANNTANATJAV;
HTVJNVAVNMVJNVBVNMVJNVTVAVBVNMV;
NKVNJAV;
NKVJJAV;
NKVRJBJAV;
NKVRJCJAV;
NKVRJDJAV;
NKVRJJJAV;
NKVRJANTJAV;
NKVRJANNTJAV;
NKVRJANNNTJAV;
HNKVBVJAVNKVRJBVJAV;
HNKVTVBVCVJAVNKVKBVCVJAV;
HNKVTVBVCVJAVNKVOBVCVJAV;
HNKVTVBVCVJAVNKVRBVCVJAV;
QNKVPABJAV;
QNKVMABJAV;
QNKVMNCJAV;
QNKVPGFJAV.

26. NKVJAVJOV;
NKVMNAJOV;
HNKVMNBVJAVNKVBVJOV;
NKVPABJOV;
HNKVTVAVBVJOVNKVMAVBVJOV;
HTVNKVTVAVBVJOVQLNBVNKVPAVBVJOV;
QNKVPANJAV;
QNKVPANJOV;
SLOPAVBVPCVDVPOAVCVOBVDV;
SHUOAVDVOBVCVUPAVBVPACVDV.QLOBVDVN;
HNKVAVJAVUMNAAV;

27. HTVHAVBVHBVAVHVAVBV;
HHVAVBVTVHAVBVHBVAV;
HVHVAVBVTVHAVBVHBVAV.

28. HLRGVBCQNKVGVJOV;
HLRGVBBQNKVGVJOV;
HLRGVBEQNKVGVJOV;
HLRGVBENKVGVJEV;
NKVJOVJEV;
NKVJAVJEV;
HLRGVBMNAQNKVGVJEV.

29. NKVMARBMNAJBV;
NKVMARCMNAJBV;
NKVMARDMNAJBV;
NKVMARNVMNAJBV;
LNLVJBVA;
NKVPAAJCV;
NKVPABJCV;
NKVPACJCV;
NKVPANVJCV;
LNLVJCVN;
NKVRMAPABBJBV;
NKVRMAPACCJBV;
NKVRMAPADDJBV;
NKVRMAPARJANNTRJANNTJBV;
NKVRMAPANVNVJBV;
NKVNLVJBVJEV;

Made in the USA
Las Vegas, NV
13 January 2024

84308847R00313